SIMULATION, KNOWLEDGE-BASED COMPUTING, AND FUZZY STATISTICS

VAN NOSTRAND REINHOLD ELECTRICAL/COMPUTER SCIENCE AND ENGINEERING SERIES
Series Editor: Sanjit Mitra

Handbook of Electronic Design and Analysis Procedures Using Programmable Calculators, B. K. Murdock
Compiler Design and Construction, A. B. Pyster
Sinusoidal Analysis and Modeling of Weakly Nonlinear Circuits, D. D. Weiner and J. F. Spina
Applied Multidimensional Systems Theory, N. K. Bose
Microwave Semiconductor Engineering, J. F. White
Introduction to Quartz Crystal Unit Design, V. E. Bottom
Digital Image Processing, W. B. Green
Software Testing Techniques, B. Beizer
Light Transmission Optics, Second Edition, D. Marcuse
Real Time Computing, edited by D. Mellichamp
Hardware and Software Concepts in VSLI, edited by G. Rabbat
Modeling and Identification of Dynamic Systems, N. K. Sinha and B. Kuszta
Computer Methods for Circuit Analysis and Design, J. Vlach and K. Singhai
Handbook of Software Engineering, edited by C. R. Vick and C. V. Ramamoorthy
Switched Capacitor Circuits, P. E. Allen and E. Sanchez-Sinencio
Software System Testing and Quality Assurance, B. Beizer
Modern DC-to-DC Switchmode Power Convertor Circuits, R. P. Severns and G. E. Bloom
Estimation and Optimum Control of Systems, T. E. Elbert
Digital Telephony, B. Keiser and E. Strange
Data Compression Techniques and Applications, T. J. Lynch
Digital Transmission Systems, D. R. Smith
Flat-Panel Displays and CRTs, edited by L. E. Tannas, Jr.
Microcomputer Systems, I. Flores and C. Terry
Computer Architecture, Third Edition, C. Foster and T. Iberall
Radiowave Propagation in Space Communications Systems, L. J. Ippolito, Jr.
Semiconductor Power Electronics, R. G. Hoft
Long-Wavelength Semiconductor Lasers, G. P. Agrawal and N. K. Dutta
Applied Reliability, P. A. Tobias and D. C. Trindade
Time Domain Measurements in Electromagnetics, edited by E. K. Miller
Digital Communications, I. Korn
Understanding Antennas for Radar, Communications, and Avionics, B. Rulf and G. A. Robertshaw

Simulation, Knowledge-Based Computing, and Fuzzy Statistics

Constantin V. Negoita

Dan Ralescu

VAN NOSTRAND REINHOLD COMPANY
New York

Copyright © 1987 by **Van Nostrand Reinhold Company Inc.**
Library of Congress Catalog Card Number 87-8249
ISBN 0-442-26923-4

All rights reserved. No part of this work covered by the copyright hereon
may be reproduced or used in any form or by any means—graphic,
electronic, or mechanical, including photocopying, recording, taping, or
information storage and retrieval systems—without written permission of
the publisher.

Printed in the United States of America

Van Nostrand Reinhold Company Inc.
115 Fifth Avenue
New York, New York 10003

Van Nostrand Reinhold Company Limited
Molly Millars Lane
Wokingham, Berkshire RG11 2PY, England

Van Nostrand Reinhold
480 La Trobe Street
Melbourne, Victoria 3000, Australia

Macmillan of Canada
Division of Canada Publishing Corporation
164 Commander Boulevard
Agincourt, Ontario M1S 3C7, Canada

16 15 14 13 12 11 10 9 8 7 6 5 4 3 2 1

Library of Congress Cataloging-in-Publication Data
Negoiṭă, C. V. (Constantin Virgil)
  Simulation, knowledge-based computing, and fuzzy statistics.
  Bibliography: p.
  Includes index.
  1. Computer simulation. 2. Expert systems (Computer
science) 3. Fuzzy sets. I. Ralescu, D. A. II. Title.
QA76.9.C65N45 1987   006.3'3   87-8249
ISBN 0-442-26923-4

CONTENTS

Preface / ix

INTRODUCTION   1

Systems, Simulation, and Control / 1
Knowledge-Based Controllers / 4
The Semantic Approach in Knowledge Engineering / 6
Knowledge Acquisition versus System Analysis / 8
The Human Management of Systems / 11
Fuzzy Sets and Systems / 15
Readings / 18

1: SIMULATION MODELS   27

    Complex Systems Engineering / 27
    Models versus Programs / 29
    Realistic Models Complicate the Programs / 32
    The Theorem of Representation Simplifies the Programs / 37
    Models in Relational Data Bases / 39
    Readings / 42

2: LINGUISTIC STRATEGIES   51

    The Rotary Cement Kiln Control / 51
    Aggregate Production Planning / 53
    An Automatic Train Operation / 55
    Investment Decisions / 58
    Inference Rules and Fuzzy Statistics / 61
    Readings / 66

3: CONCEPTS OF FUZZY SET THEORY   79

Fuzzy Sets and Fuzzy Relations / 79
Algebraic Theories of Fuzzy Sets / 84
Representation of Fuzzy Sets / 89
Readings / 94

4: FUZZY RANDOM VARIABLES   99

Random Sets and Their Expected Values / 100
Fuzzy Random Variables / 107
Expected Value of a Fuzzy Random Variable / 111

5: LIMIT THEOREMS FOR FUZZY RANDOM VARIABLES   115

Limit Theorems for Random Sets / 116
Strong Law of Large Numbers for Fuzzy Random Variables / 119
Central Limit Theorem for Fuzzy Random Variables / 121

6: FUZZY SET-VALUED MEASURES   125

Set-Valued Measures / 126
Fuzzy Set-Valued Measures / 130
Strong Law of Large Numbers and Fuzzy Set-Valued Measures / 133

7: RELATIONSHIPS BETWEEN FUZZY RANDOM VARIABLES AND FUZZY SET-VALUED MEASURES   137

Radon-Nikodym Theorem for Fuzzy Set-Valued Measures / 138
Conditional Expectation of a Fuzzy Random Variable / 139
Fuzzy Martingales / 141

8: THE BAYES FORMULA FOR FUZZY PROBABILITIES 143

The Bayes Theorem for Set-Valued Measures / 143
The Concept of Independence / 145
The Bayes Theorem for Fuzzy Set-Valued Measures / 146

BIBLIOGRAPHY 149

INDEX 153

PREFACE

Any assemblage of things forming a regular and connected whole is a system. This definition is vague, but confidence in the truth of a vague assertion may be justified just because of its vagueness. System analysts aim to understand how an existing or proposed system operates. They are interested not in the physical things forming the whole, but in the plan or scheme according to which things are connected.

This book is concerned with understanding connections. The analyst looks at the system to get information. Quite often this information will disclose an unsuspected relation. This relation is a model, an explanation describing what happens. Because different analysts are interested in different aspects of a system, the system can have different models. Also, as understanding changes over time, the same aspect can lead to different models.

Given a model that mirrors a system, it is possible to derive information about the system's behavior just by studying the model's behavior. Usually this is done by computers. Depending on the method of computation, it is a *simulation program* or a *knowledge-based system*.

In simulation, the analyst internalizes the model. Internalization is appropriation. The analyst transforms the system from structures of the objective world into structures of the subjective consciousness. With the model in mind, the analyst embeds it in programs where specifications of what must be done are established.

In knowledge-based systems, the model is internalized inside the computer. The analyst transforms the system from structures of the objective world into new structures of the objective computer. With the model in its knowledge base, the machine embeds it in programs. The relationships defining the model, scattered throughout a stored knowledge base, can be used as logical sequences of steps defining an inference engine.

In simulation programs, instructions are logical sequences of steps furnished by humans who understand the problem. In knowledge-based systems, the machine understands questions and answers them based on a world view stored by the analyst.

At the beginning, system analysis looked toward the controlled system. Recently, this orientation has been changed. System analysis now looks at the controller: Looking at the controlled system is the old paradigm of physical sciences; looking at the controller is the new paradigm of artificial intelligence. Looking at the controlled system involved building mathematical models embedded in hard-coded programs. Looking at the controller involves building linguistic models embedded in knowledge-based, state-driven systems.

The main purpose of this book is to link simulation with knowledge engineering and to show how the second has influenced the first, in order to present the relationship between natural language and model building.

The impact of knowledge engineering upon simulation is illustrated by qualitative analysis, which led to verbal models whose variables are words rather than numbers. Not only can the values of the variables be linguistic rather than numeric, but causal relationships between variables can be formulated verbally rather than mathematically. The moral is that linguistic models, which dominate the soft sciences and heuristic knowledge in general, can be handled now by computers if the meaning of the words (models of the world) is internalized in a knowledge base.

The knowledge kept in a knowledge base is procedural. Pieces of knowledge in the form of rules can be chained in an inference process. In other words, from some premises one can obtain a conclusion. If the internalized premises (the rules) match the external premises (facts furnished by users), the inference process is started and the machine exhibits intelligence.

As in other areas of computer science, progress in knowledge engineering has been made by trial and error. Special cases were studied in depth, and the knowledge gained was purified and unified in what seemed to be a doctrine.

The classical inference mechanism used in practical knowledge-based systems, say before 1980, dealt with certainty factors and probabilities. Since the 1970s, the pioneers who built the first expert systems have focused primarily on these two concepts. Soon they realized that a theory of evidence in terms of probabilities would be less than ideal if the probabilities invoked are impossible or difficult to determine.

At that stage, the idea of a fuzzy variable was not precisely defined, although the term was used. Some doubted whether arguments were valid when operating directly on fuzzy variables.

With the appearance of fuzzy statistics, it became clear that fuzzy variables could be handled rigorously. This opened the field of fuzzy systems for serious study. The starting point of this study was to extend the classical probabilistic approach. What happens, for example, with probabilities that take linguistic values? Do they still have desirable qualities similar to those possessed by numerical probabilities?

Most of this book is an attempt to answer these questions. Such answers certainly constitute a legitimate mathematical activity on its own, but that has not been our only purpose. Instead, we have been motivated especially by the current interest in knowledge engineering: the manipulation of imprecise and subjective knowledge.

The major tool in our attempt is *the theorem of representation*. A fuzzy set can be represented by a family of crisp sets, which we call level sets.

Beginning with our 1975 monograph (*Applications of Fuzzy Sets to Systems Analysis*, Wiley), many mathematicians and practitioners have studied and applied this theorem. This book is another proof of its potential.

CONSTANTIN V. NEGOITA
DAN RALESCU

SIMULATION, KNOWLEDGE-BASED COMPUTING, AND FUZZY STATISTICS

INTRODUCTION

SYSTEMS, SIMULATION, AND CONTROL

We use the word "system" almost everyday. We ourselves are biological systems, shaped by educational systems, coping with economic systems, and caught up in political systems. All of these systems have something in common: They consist of interrelated subsystems, each of which contains its own interacting subsubsystems that work together to convert inputs to outputs. This is one reason why we can define a system as a relation between inputs and outputs.

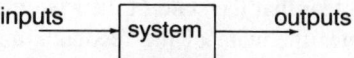

No matter how simple or complex the system, its successful use depends on an understanding of its structure. This is the main task of system analysis. The job of the analyst is to study how an organized whole can process the inputs. The analyst wants to know the character of the system in order to forecast its future. Destiny means evolution over time, and this evolution is governed by the character of the system—that is, by its structure.

Usually, we can model the structure by fixing our attention on the state of the system. The state is defined as the minimum information required to describe the system's condition in such a way that, if the inputs are known, then the condition at any time is completely determined.

A structure-determined function known as the *dynamics* describes the change of state of the system over time. When this dynamics is known, we can simulate any possible evolution in time by computing transitions from one state to another. The dynamics, in other words, is simply a state equation specifying that the state of the system at any moment is a function of the state and the inputs in the previous moment.

Therefore, it seems that one of the analyst's jobs could be to find the state equation, or model. The next step is to transform the model into one or more statements as part of a program operating on data. These data are numbers.

Sometimes we have to cope with nonnumerical data. Linguistic values like "many" or "high" are frequently used in scientific descriptions. Facing this new complication, the analyst has to translate words into numbers. This translation is known as the *semantic* approach.

The analyst may also want to recommend those inputs that are supposed to determine a specified evolution in time. Given a structure, we may want to know how to improve the system's behavior—how to find desirable controls in order to achieve some objectives. Over a long time, a theory of control was developed, and its highlights were considered as an acceptable framework for studying predetermined behaviors.

Future projections or imagined trajectories can be simulated by processing data. Processing vast amounts of such data by hand can be time consuming and costly, hence the use of computers.

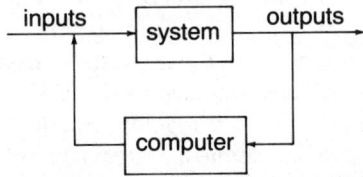

For the analyst, the controller is just a new system whose state is changing in time. It is perfectly clear that the state of the system determines its output, which in turn determines the state of the feedback and, therefore, the system's input adjustment. Again we can say that the controller's state at any moment is a function of its state in a previous moment and the system's output in that previous moment, thus defining the controller's state equation. The whole *feedback loop*, input-system-output-controller, has a very special behavior: It manifests *purpose*.

More than 40 years ago (Rosenblueth, Wiener, and Bigelow, 1943) the world of science was shocked to find out that in nature such teleological systems exist. So shocked, in fact, that they reinvented an almost forgotten science—*cybernetics*, the interdisciplinary science of communication and control. They observed that the feedback loop can be identified in society, men, animals, and machines. In all these realms purposeful behavior is directed toward a final condition in which the behaver reaches a predetermined relationship with respect to another entity. Without too much hesitation, they concluded that all purposeful behavior needs feedback; that is, the behavior of a system is controlled by the margin of error at which the system stands with reference to a *goal*. Consequently, teleology—the science that treats the end or design for which things were created—was defined as the science of purpose controlled by feedback.

If the principles of cybernetics are applied to labor-saving machines and methods, the result is *automation*: the technique of making an industrial process or system operate automatically, and more recently, involving electronic devices for controlling processes or systems.

In the 1970s, automation, with its emphasis on controllers, brought the internal model principle to the fore. According to this principle, a control system needs feedback and must incorporate in the feedback loop a model of the dynamics of the external world. Internalizing a model of the structure of the controlled system and of the environment affecting this controlled system was the way to manifest purpose.

If this function is known, then everything seems to be in order; however, usually it is not. An important aspect of system analysis so far has been the development of a unique model, a unique description. But our inability to make precise descriptions forces us to deal with partial descriptions. If we explore the mode of cognition as a procedure for describing complex systems and how natural language is involved in this process, we see that humans cope with complexity by synthesizing partial representations and that learning implies conceptualization, which is concerned with getting a feel for the whole.

To make this statement more understandable, we recall the story of several blind men who were asked to describe an elephant. One, touching his legs, described the animal as a pillar. Another, touching his ear, described him as a fan. A third, touching his trunk, described the animal as a water pipe. In the search for unity behind diversity, conflicting representations reflect the nature not of things but of the perceiving mind.

Therefore, Aristotelian logic, based on the law of the excluded middle (x cannot be A and non-A at the same time), has to be replaced by a dialectical logic, assuming that A and non-A do not exclude each other as the predicates of x.

If *dialectic* means nothing else than a critical analysis of mental processes, and *identity* is nothing given or defined but something to be continually achieved, then the whole theory of teleological systems can be restructured on the idea of conflict resolution. The existence of a conflict at one level of description generates a synthesis at a higher level. This movement from one level to another can be described as the dynamics of the system of representation.

In this book, the definition of a purposeful system is enlarged by adding the idea of purpose controlled by *pullback*, when pullback is understood as a special movement in the structure of partial representations. If partial representations are organized as a knowledge base, then the operation of a knowledge-based system can be described by such movements.

The *intelligent* behavior of a knowledge-based system can be explained as the result of *internalizing* a knowledge base, and, therefore the structure of facts and, therefore, the properties of this structure.

KNOWLEDGE-BASED CONTROLLERS

The idea of a controller is based on system *observability*. A controller observes the output of the system and acts accordingly. The observed output of the system becomes the premise, and the action becomes a conclusion of an IF-THEN rule:

IF output THEN input

One possible way to identify the states of the system is to consider the equivalence classes of the set of all possible inputs. For instance, input A is equivalent to input B if the output is the same. The system is *observable* whenever two distinct states yield observably different responses; that is, for two states there is at least one input to which they react differently. Observability is a vague predicate. We commonly distinguish outputs that are similar from those that are not.

A similarity relation has the following properties:

1. Input A is similar to input A.
2. Input A is similar to input B means input B is similar to input A.
3. Input A is similar to input C means input A is similar to input B AND input B is similar to input C.

A *congruence* is a special similarity relation satisfying the property

4. Input A followed by input Z is similar to input Z followed by input B means that input A is similar to input B.

When two inputs follow each other, they are said to be *concatenated*.

With this approach we can consider simultaneously a family of IF-THEN rules, each corresponding to a level of similarity. If this level of similarity is measured with numbers belonging to the unit interval, a family of controllers can be described as a vague statement:

IF vague output THEN vague input

where "vague" now has a precise meaning. We say

vague input: input → [0, 1]
vague output: output → [0, 1]

Therefore, a vague input or output is a function defined on input or output with values in [0, 1]. The feedback loop becomes

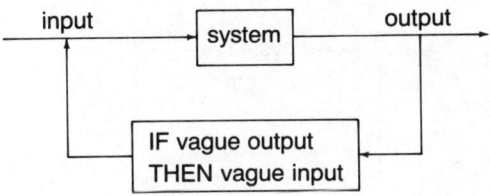

For simplification we replace the IF-THEN rule by an arrow. Therefore, IF A THEN B becomes $A \to B$.

Usually, the controller is composed of a number of rules:

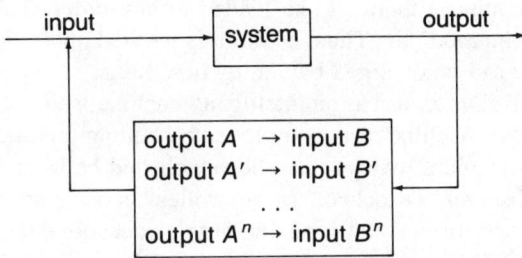

The key to the controller intelligence is that input is taken either from an *aggregation* of the contributions of all the rules or from a *selection* of the *best* rule. All the rules together can form a knowledge base internalizing the knowledge of a human operator.

The rules can also be activated sequentially in a chain:

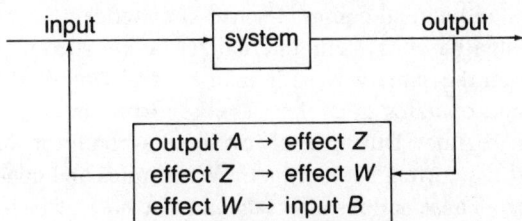

The chaining of the rules is a line of reasoning or, in other words, an inference mechanism that can be thought of as an executable image of a knowledge base, ready to be accessed by a driver. The major part of running a knowledge-based system is getting a driver program to load and use the inference mechanism. No predefined structure determines the relationship among rules. Rules establish hooks to each other as they are loaded. A realistic situation could be

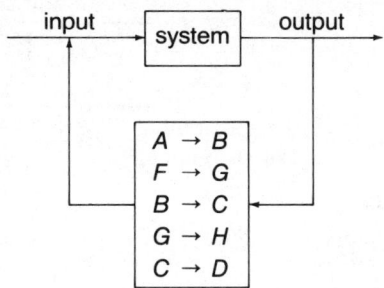

This characteristic yields a fascinating property of inference mechanisms: The rules that compose them can be loaded in any order. The knowledge-based is said to be modular. Thus, as more is learned about the system, the knowledge base can be enlarged by adding new rules.

Now the controller is seen as an intelligent machine, self-building its lines of reasoning. This machine can be improved by adding facilities to pose and answer questions, and a dialogue can be established between the controller and a human observer. Detached, the controller becomes an *expert system*, an information system that can pose and answer questions relating to information borrowed from human experts and stored in a knowledge base.

THE SEMANTIC APPROACH IN KNOWLEDGE ENGINEERING

The discipline of knowledge engineering has emerged from the proliferation of knowledge-based systems. The goal of any knowledge-based system is to transfer, utilize, and extend common-sense knowledge.

Any knowledge-based system internalizes a knowledge base to solve problems in much the same way as human experts would. It contains task-oriented rules and operates in cycles. The key to its intelligent behavior is the state-driven regime. During each cycle, the condition of each rule is matched against the current state of facts. When rules and conditions match, actions are taken. These actions affect the current state of facts, making new rules match.

Building a knowledge base means transferring expertise from human experts, who use natural language dominated by vagueness. In our use of natural language there is no determined point at which a transition occurs from a clear case to a borderline case. Vague expressions occur due to the existence of experienced continua of qualities and the absence of fixed habits of discrimination between segments of such continua.

We cannot draw a boundary, except arbitrarily, between cases when a word applies and when it does not, simply because a continuum exists that makes it impossible to do so satisfactorily.

An essential characteristic of any word is that the boundaries of its applicability are not fixed. The question of truth and falsity is not only undecided but undecidable. Any word covers a multitude of real objects, situations, or predicates. Each such object, situation, or predicate is covered by the word with a degree of intensity. Indeed, we seem to relish this linguistic looseness and even compound it by talking about "a few" or "many." But how many are a few?

Vagueness is a matter of context and, therefore, of subjective evaluation. But when we turn to computers to solve problems, such undecidability invites trouble. Human expertise used to be almost totally beyond the range of machines because it usually involves vagueness.

How, for instance, could a computer tell when the temperature of a cement kiln is "high" or when its fuel feed rate is "OK"? How could a computer tell when a patient is in a "good" state, or understand a doctor's statement that "acute" pyelonephritis "usually" presents "bladder irritation"? There is only one way: It must understand these words as human experts would. To understand means to know the meaning of. Meaning is given only by persons. Meaning is what a person intends, aims, or purposes. Therefore, when translating the sense of the words from natural languages to machine languages, we have to capture the intent of the person who handles the words.

In a computer, information is represented, stored, and processed in binary form—a coding scheme of binary digits. The computer's memory is simply a set of cells containing these binary digits. Because of our coding scheme, the same pattern of binary digits in a memory cell can be interpreted as a number, as an alphanumeric character, or even as an instruction in a program. We tell the computer how to interpret that pattern through the program.

More than that, we can define and name complete subprograms or miniprograms, which do a specific task, and use them by putting that name in our main program. In this way we give meaning to that name.

Let us use this approach to represent the meaning of the words we use in everyday communication. Consider the word "old." This word can be *defined* as a continuum of possible choices, according to the universe of possible human ages. Clearly, someone over 80 is old, so the degree of oldness for 80 or greater could be on a scale from 0 to 1. It is not clear that a 60-year-old is old.

We evaluate the degree of oldness at age 60 as 0.7. Rather than saying that a 60-year-old is old or not old, one can say that the individual is *partially old*. In this way, the vagueness of the term "old" can be captured numerically in a table:

Old

Age	Degree
30	0.1
40	0.2
50	0.4
60	0.7
70	0.8
80	1.0

The meaning has been captured by a data structure, and a vocabulary becomes a semantic system. Therefore, a semantic system is a knowledge base of meanings translated as data structures.

Internalizing a semantic system as a knowledge base of meanings, the computer becomes intelligent, not only because it contains all these meanings but because by accepting this representation of meaning, the properties of the structure of such tables are internalized as well. By accepting a list of tables as vocabulary, the computer implicitly has the inference engine to combine them in sentences. All the operations can be internalized as miniprograms, and the inference takes place automatically by a procedure invisible to the user.

Consider a data base with a semantic system and a query for retrieval formulated linguistically as

Find the best companies with high profit.

Every word has a meaning, and high-level retrieval is precisely a form of inference. This is the case for *expert data bases*. With the vocabulary embedded in a high-level language accepted by the machine, any text in natural language can be considered as a source program statement. The interface between the machine and the user is a natural one, that of the natural language.

KNOWLEDGE ACQUISITION VERSUS SYSTEM ANALYSIS

The term "system analysis" has had a somewhat adventurous history. It was originally employed in the wake of information systems to denote the first phase in a process starting with the identification of problems. System analysis examines an existing system with the express goal of computerizing the information channels. The system analyst was considered as an architect, who began the cycle by interviewing the user. After defining the user needs, the analyst draws tentative plans for a suitable *data base*.

In management sciences "system analysis" denotes the identification of models (relations between data). So after identifying the user needs, the analyst draws tentative plans for a suitable *models base*.

In artificial intelligence the term was replaced with "knowledge acquisition" to protect it from any mathematical or data base connotation.

If this new term is preferred, as it probably is today, then the question of whether the secularizing potency of artificial intelligence was a novum or had its roots in earlier elements of the computer-based systems tradition inevitably suggests itself. We contend that the second answer is correct. The roots of knowledge acquisition can be found in early sources. To appreciate this position, one must see an information system in the context of the cultures amid which it sprang up and against which it defined itself.

The computer must be thought of as an extension, as with any other product, of the industrial society, like cameras or cranes, which represent extensions of sensory or physical capabilities. Because of the unfriendliness of computers, their complexity is sometimes perceived as a barrier. Today between the user and the computer there is an interface, the analyst and the programmer. Any intelligent action is taken by these interfaces.

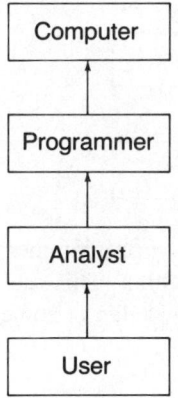

Any traditional computer program directs the machine to access data, but the decisions about how to process those data are invariably hard coded in the language of the program and stored in memory during program execution. These decisions are made by a human programmer with the knowledge to make them.

The programmer writes the program, and the computer works for him.

10 INTRODUCTION

Early computers were programmed in machine language. Programmers had to remember the codes of the various computer operations. This was a tedious and error-prone process. And because the resulting programs were difficult to read and modify, assembly languages were developed. An instruction in assembly language is an easy-to-remember form called a mnemonic. This step was in the right direction, but the programmer was still forced to think in terms of individual machine instructions. Because humans solve problems and communicate in natural languages, higher programming languages were developed to be closer to human thought. In these higher-level languages, names were used for subprograms, procedures, and functions. These languages were extensible; that is, they could easily be extended to do more things. The language permits the programmer to internalize more knowledge about processing data. Thus the programmer becomes more intelligent.

Knowledge engineering is a discipline devoted to integrating knowledge in computer systems. Consider the semantic system, which is a vocabulary installed into the machine

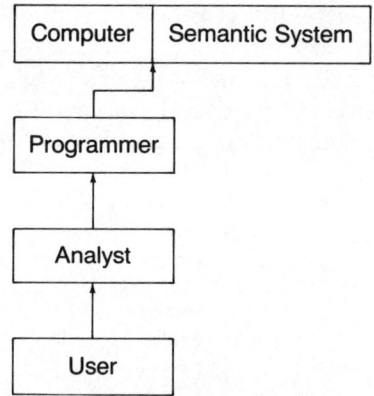

The next step is to get rid of the programmer and the analyst and to install their knowledge inside the machine. This can be done if a knowledge engineer has already extracted knowledge at some previous time. The situation becomes

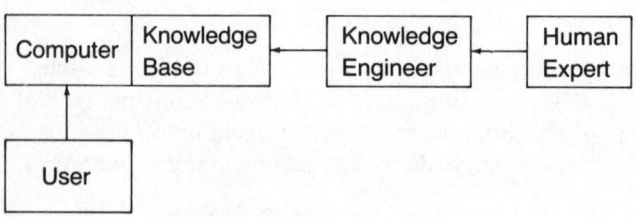

The distinctive characteristic of any such system is that its processes are state driven rather than hard coded. Decisions about how to process data are part of the knowledge of the system. In other words, an intelligent system writes its own program. Internalizing procedural knowledge as a model of the world, the machine becomes intelligent. At least, this is the way the user sees it. The user can have a direct dialogue with the intelligent machine, without any interface, because the intelligent machine understands his or her questions. However, between the human expert and the machine there is still an interface, the knowledge engineer.

The next possible step is to internalize the knowledge engineer. This could be done by the semantic system

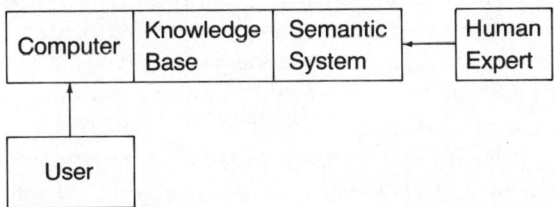

A semantic system can take the place of a knowledge engineer, accepting statements in natural languages and complementing the normal knowledge base that stores the expertise of the expert. It is not too difficult to see that this configuration can have far-reaching consequences for the power of the knowledge-based system. The semantic system is storing knowledge about knowledge, or metaknowledge, supplementing the inference power of the knowledge base. The expert system becomes a program that can handle more sophisticated humanlike judgments.

The contemporary situation of knowledge-based systems is thus characterized by a progressive internalization of knowledge. Internally, the knowledge base is no longer just a set of rules. The operation of these rules is now dominated by the logic internalized in the structure of the facts over which the rules were defined. Externally, the knowledge-based system deals with both builders and users in the same way. For answering questions or acquiring knowledge, in all these aspects of their mission, knowledge-based systems are compelled to seek results by methods that are, of necessity, very similar to those employed by humans.

THE HUMAN MANAGEMENT OF SYSTEMS

Of all the fields metamorphosed by the computer, none has changed more than management. There is an almost quantum differential between the traditional management information system of the 1960s and the decision support system of the 1980s. The nature of this paradigm shift has been widely discussed among the leading thinkers in the field and has generated a considerable volume of literature, scholarly or otherwise.

The prevalent paradigm of decision support organization, with scientific overtones, has demonstrated its weaknesses, however, and the greatest of these has been a lack of concern for the elements of *human* management. In their quest for what they have assumed to be scientific precision, too many contemporary management scientists have ignored the facts that cultures are idiosyncratic or pluralistic and that their imprint on the decision support system can include more than mathematical models acting on data bases.

Of late, as the more publicized paradigm shift toward relational data bases methodologies was becoming ever more firmly entrenched, a quieter revolution has been generated. This revolution has involved an increasing focus on perspectives that can best be termed "knowledge based."

In the periodic ritual of self-examination to which most academic disciplines are given, much has been made of the *data base approach*, but very little has been written of the more recent *knowledge-based approach*. Some recent works have been pointed to by some as indicative of the new growing "human" emphasis. Collectively, these works have been interpreted by many as criticism (implicit or explicit) of some of the fundamental assumptions in the scientific management, which has seemed increasingly preoccupied (if not enamored) with the formulation, generation, and testing of system models, models describing the system to be controlled.

Those who have espoused a more human doctrine in the doing and the writing of management for the most part recognized that there is another, perhaps better, possibility—trying to represent the human controller. This possibility led to knowledge bases and their theory.

Suppose we want to set up a data base describing the structure of a bicycle, using the vocabulary bicycle, wheel, hub. If the data base contains the phrases

> wheel   part of   bicycle
> hub   part of   bicycle

then the query

> Is wheel part of bicycle?

which asks if the sentence is in the data base, has a positive answer. The query

> Which is part of bicycle?

finds all the instances of facts that can be confirmed (in our example, wheel and hub).

The bicycle data base used the single relation "part of" to describe the structure of the bicycle. For this reason the query

Is hub part of bicycle?

gets a negative answer, even though a hub is of a bicycle, because it is a part of the wheel, which is part of a bicycle.

To answer the question, the intelligent human outside the data base must define another relation, "indirect part of," by a recursive (i.e., coming back to itself) definition:

1. x indirect part of y if x part of y.
2. x indirect part of y if z part of y AND x indirect part of z.

The logical reading of this definition is quite simple: To solve a condition of the form B, solve the condition A; that is,

$$A \rightarrow B$$

Formulating this implication, the external human operator behaves in an intelligent way. If the operator expects to receive many similar questions, the best thing is to introduce the rule in the data base. With this piece of knowledge inside, the data base becomes an *expert data base* because it can display a line of reasoning.

A normal step further is to accumulate such rules in the data base and eventually to *ask for facts* when they are needed. Suppose we are interested in using the data base to help us to repair a bicycle. Before we can use it, we should perhaps get hold of a bicycle repair expert to tell us what symptoms indicate a disease in some part of the bicycle by giving us some facts of the form

Problem indicates (fault in part)

as in the sentence

Flat tire indicates (puncture in wheel)

Consider the pair of rules

1. x possible fault in y if z indirect part of y AND x indicates (x in z) AND x is reported.
2. x is reported if (x a problem) is told.

A query like

Find all possible faults

sets up an interaction (dialogue) of the form

> Machine: *x* a problem?
> User: Flat tire
> Machine: Puncture possible fault

with the meaning that the expert's data are accessed in a particular, systematic way.*

The conclusion of the previous example is that the *old paradigm* of a data base,

> Human operator's rules   *exploit*   facts in data bases

is replaced by the *new paradigm* of a knowledge base,

> Human operator's facts   *trigger*   rules in knowledge bases

The first paradigm characterizes operations research. The second characterizes artificial intelligence.

It is well known from the 1960s that optimal control theory provides a powerful method of deriving rules that enable a dynamic system to be held to a preferred path. Optimal control theory thus appears to be eminently suitable for assisting managers to improve the performance of their controlled systems. According to the broad concept of regulation and tracking, the object to be controlled admits as inputs control signals supplied by a controller on the basis of information about the specific response to the control and disturbance signals currently being processed.

Central to a classical control theory is the idea of a model that describes interactions between the system variables. Although a multitude of mathematical models exist in the literature, implementation of any of these models is practically nonexistent. Most managers have decided to use their own heuristic or intuitive decision rules, which do not guarantee mathematical optimality.

Humans can receive, perceive, and remember a limited amount of information. Verbal recoding is the human way of repackaging material into a few chunks rich in information. Human controllers use these chunks of knowledge, which can be stored in knowledge bases. This trend shows in application the advantage of a broader, more explicit, cultural approach for the methodological and empirical inquiries in management science.

The knowledge-based approach should be most useful to anyone trying to define management in a strictly traditional way from operations research texts constrained by mathematical language. After all, the recent development of expert systems is proof that a linguistic model is as good as a mathematical one.

---

*K. L. Clark and F. G. McDale, *Micro-PROLOG: Programming in Logic*, Prentice-Hall, Englewood Cliffs, NJ, 1984.

FUZZY SETS AND SYSTEMS

We said that a fuzzy set can be a good model for describing the meaning of a vague word. We also said that this fuzzy set is a function defined on a universe of discourse with values in an ordered set, usually the unit interval:

$$\text{Fuzzy set: Set} \to [0, 1]$$

Some commentators say that the idea of a fuzzy set derives from mathematics. Nothing could be more mistaken. The idea comes from engineering. It was born to solve very practical problems in pattern recognition. The first papers demonstrate this.

Theoretically, such problems were previously described by set theory. One way to describe a set is by a characteristic function defined on a universe of discourse with values in a set with only two elements:

$$\text{Set: Universe} \to \{0, 1\}$$

Zadeh (1965) suggested using the unit interval $[0, 1]$ instead of the set $\{0, 1\}$, thus making the membership a matter of degree. The first impulse was to say that a fuzzy set generalizes a set. Unfortunately, it is not so easy to prove such a statement.

At almost the same time, mathematicians working on modern algebra, completely unaware of the innovations made in engineering, tried to investigate the idea of a generalized set. A set is a collection of things sharing a common feature. The power of set theory is that it allows thinkers to deal with many as one, without getting bogged down in individual differences. When defining the set of mountains, nobody cares about their heights. All mountains are in the set of mountains because each has the property of a mountain.

If we care about heights, then the first observation would be that not all mountains are equal. All mountains are mountains, but there are mountains more mountainous than others. Thus, the collection of mountains is not exactly a set where all the elements are the same, but a generalized set where not all the elements are the same. We say that the elements are not equal. What these mathematicians did was to fuzzify the identity, a vital property not to be neglected in mathematics. Therefore, in a generalized set the elements have a degree of existence.

In the last two decades an interesting theory of generalized sets has been developed independently of the theory of fuzzy sets. Only very recently, have there been some atempts to see how these two theories can merge. The reason for such attempts was again very practical. The theory of generalized sets is at the heart of modern symbolic logic. The idea of a fuzzy set is at the heart of linguistic modeling. Practically, any predicate can be modeled as a fuzzy set. Therefore, the combination of predicates is a combination of fuzzy sets. This combination is achieved by the logic operators AND, OR, NOT, and their definitions influence the result of the combination. Therefore, a good definition is vital. But what is a good definition?

One of the most important results of the theory of generalized sets is that any logic is internalized in a structure of generalized sets, and, in particular, a multivalued logic is a consequence of the existence of such a structure. The practical conclusion is that the logic operators depend on the type of knowledge representation. If we can show any relation between fuzzy sets and generalized sets, we can hope to borrow the conclusion.

One possibility for moving from the structure of fuzzy sets to the structure of generalized sets is to represent a fuzzy set as a family of crisp level sets. This representation, proved in 1973, led to a mathematical theory of fuzzy sets with very practical results, one of which is fuzzy arithmetic.

Another result was the development of a theory of fuzzy systems. The idea of a system, the abstract system, along with increasing concern for what is synthetic and global as opposed to analytic and local, dominated the scientific community in the last century.

Modern system theory developed as a mathematical theory. Central to system theory are the notions of behavior and connection, and the first problem encountered is finding the behavior of a connection from the behaviors of the things connected.

The mathematical system theory extensively promoted during the 1960s has lately come under a barrage of criticism, and a new movement "back to the basics," has come to the fore.

People who have attempted to analyze complex problems from a systems point of view have frequently remarked that the symbolic language used in building a model is too poor to convey all the nuances a person can carry in mind.

The struggle to reconcile mathematics and the soft sciences has its parallel in automatic control, also dominated by pure mathematics. Among control theorists there have emerged advocates of the need to develop a fuzzy approach. Taking a retrospective view of the development of the theory of automatic control, from its beginning to the time when interest in variational techniques reached its peak, they observed that this theory is permeated with unfulfilled hopes. Many problems remain unsolved. They can be reduced, in the main, to analyzing an equation that depends on parameters. In general, simple solution properties abide in complex, difficult-to-describe, and highly irregular parameter regions. By questioning the language used to describe these regions, they observe that applied mathematics is not a second-rate mathematics, nor is it merely an application of pure mathematics to practical problems. It is a different kind of discipline that should be based on a different concept of precision.

Classical system theory is based on the theory of sets. From a fuzzy set one can try to build a theory of fuzzy systems. More than that, if we consider the whole structure of fuzzy sets, then its properties can be put into evidence and used in knowledge engineering. Their adoption as a scientific approach does much to mitigate the mathematics–nonmathematics dichotomy and improve communication across scientific disciplines.

The theory of fuzzy sets is generally thought of in one of two very different ways. When thought of by applied decision theorists, it is usually as a course in naive fuzzy logic showing how its principles can be used in various practical situations. When thought of by mathematicians, it tends to be presented as a collection of techniques useful in establishing an independent theory.

The original goal we had in writing this book was to find some middle ground. We wanted a book that discussed the more theoretical ideas and techniques in a manner that was consistently oriented toward solving problems in knowledge engineering. According to what we have said so far, by knowledge engineering we understand not only the management of production rules by symbolic manipulations but also the management of vocabularies by semantic manipulations.

This original goal was indicated by our position at the time, which can be described as basically pragmatical. In particular, it seemed crucial to emphasize how the various techniques could be used. During the writing we changed our minds. It then seemed crucial to emphasize why the various techniques should be used and why fuzzy set theory is needed at all.

There was no simple cause for this conversion. We gradually realized that things ultimately seem to make sense only when looked at from a solid theoretical viewpoint. In other words, we rediscovered the old principle that nothing is more practical than a good theory.

A compelling consideration was the wide acceptance given to the representation theorem (any fuzzy set is a family of level sets) in trying to embed fuzzy set theory in contemporary mathematics.

Contemporary mathematics is based on Cantor's set theory, the main principles of which assume that actual infinite sets exist. During the last decade, however, another view has been nurtured by Vopenka and his colleagues at Charles University in Prague.

Roughly speaking, this alternative set theory seeks to provide a foundation for mathematics based on the human observer rather than, as in Cantor's set theory, on idealizations transcending experience. Actual infinite sets are no longer admitted; all sets are finite in the traditional sense. What takes the place of actual infinity is the phenomenon of infinity as we experience it when observing large incomprehensible finite sets.

To illustrate what is intended, we use a variant of Hilbert's hotel with a denumerable infinity of rooms in which a new guest can be accommodated, even when all rooms are occupied, by moving the guest in room n to room

$n + 1$, thereby vacating the first room. In Vopenka's hotel there are only 1000 rooms, all occupied, but a new guest is accommodated by the same strategy: Put him in room 1, move its occupant to room 2, and so on. Since guests are moved successively, the process will not be finished in one day, so each guest will be accommodated for almost the entire day. The set of rooms contains a subcollection of all rooms to which guests are potentially moved that behaves somewhat like the countable set in Cantor's theory.

An obvious objection to this example is that it depends crucially on the lack of precision of the key concepts "potentially moved." What Vopenka is proposing is that this is what infinity is all about.

As in traditional theory, classes are collections of sets. Some classes are sets, but others, called proper classes, are not. Again, as in traditional theory, some classes are proper because they are too big (as the class of all sets). However, and this is the key idea, classes can also be proper because they are imprecisely defined. Proper classes of this sort are called semisets. Novak (1984) approximates a semiset as a fuzzy set. A discussion of these ideas can be found in Chapter 3.

This attempt could be easily related to one made by those who tried to embed the theory of fuzzy sets in the theory of generalized sets (topoi), developed by Lawvere and exploited by logicians, who noted that a generalized set is governed by a multi- (infinite) valued logic.* The mathematics seminar led by Ponasse at the Université Claude-Bernard in Lyon was a forum for this. Again, the theorem of representation was the tool used for embedding.

READINGS

J. Casti, 1984, System complexity, *Options* (Journal of the International Institute for Applied System Analysis) **3**:6–9.

The common notion of a complex system—that it constitutes numerous variables interacting through many feedback–feedforward loops producing surprising counterintuitive behavior—tacitly assumes that the system is viewed in a uniformly agreed upon fashion, with various measures of complexity as consequences of the manner in which the system is seen. But what one sees depends on one's viewpoint. In this essay John Casti (author of *Dynamical Systems and their Applications*, Academic Press, New York, 1977, and *Connectivity, Complexity and Catastrophe in Large-Scale Systems*, Wiley, New

---

*Writing a book containing many intuitive ideas about fuzziness poses the problem of acknowledging the originators of the ideas. Many originators become part of the folklore, and many ideas have been independently rediscovered by others. We have tried to acknowledge the originators whenever they were known to us.

York, 1979) argues for a view of system complexity as a property arising from the interactions of the system with its observer/regulator rather than as an intrinsic property of the system itself. Complexity, like beauty, is as much a property of the beholder as of the object being observed.

Despite its currently fashionable status in certain corners of the applied and theoretical system world, complexity as a concept is certainly far from a new idea. For at least as long as there have been tax laws, government, university and corporate bureaucracies, and economic forecasters, each inellectual era has pitted its theories and technology against the problem of coping with the burgeoning complexity of its social, political, economic, and technological systems. The intellectual and philosophical graveyards are littered with the corpses of these failed efforts. Casti's contention is that these failures are attributable to the fallacy of regarding complexity as an observer-independent system property, itself an attitude originating in the Cartesian mind/matter duality and its consequent philosophy of reductionism.

Casti presents a nonreductionist framework for identifying system complexity and its management. The ideas form the germ of a vastly more ambitious research effort that one might most appropriately term a "theory of models," a theory that, by and large, does not exist. A more complete characterization of complexity along the lines sketched here offers the promise of filling at least one of the stars in the almost limitless constellation we call system analysis.

At a rather general level the types of questions that such a theory of models should address include the following:

What is the relationship between a natural system and its model, which is a formal system?
How do we choose an abstract state space, and how do we operationally define an observable on this space?
What types of formal mathematical systems can be used to model a natural system?
How can we compare two different formal systems purporting to model the same natural system?
For a given natural system how do we generate new predictions from previous predictions?

O. G. Selfridge, E. L. Rissland, and M. A. Arbib (eds.), 1984, *Adaptive Control of Ill-Defined Systems*, New York: Plenum Press.

This is a fairly complete survey of significant opinions presented at a NATO Advance Research Institute held in England in 1981. Most of the papers of this contributed volume deal with the definitions of the ill-defined system.

Neville Moray, from the University of Toronto, says, for instance, that an ill-defined system is one whose state-transition matrix cannot be known because some states are inaccessible, some of the transition probabilities are inaccessible, or the matrix is not time invariant. Although many industrial processes and man-machine systems in general can be to a good approximation regarded as well defined, at least as far as the machine is concerned, the properties of the human operator are such as to make the man-machine combination of an effectively ill-defined system. What is surprising is that nonetheless human often control them rather effectively.

Richard Young, from the MRC Applied Psychology Unit, Cambridge, England, says that an essential property of an ill-defined system is that it lends itself to no simple analysis that is both correct and adequate to answer all relevant questions. Instead of this, we have to use a variety of different views, hoping to find (at last) one that is appropriate for a given purpose. If our intention is to create techniques for understanding and controlling complex and ill-defined systems, a useful viewpoint to consider is that of the operator (or user) of such a system. In contrast to other approaches that focus more on the formal analysis and technical properties of the system itself, this orientation places its main emphasis on an essentially psychological question—that of the structure and content of the knowledge needed to utilize the system effectively.

Margaret Boden, from the University of Sussex, England, says that the concept of ill-defined system collapses into triviality if it is used to refer to any system that has not yet been well defined. We might instead take it to mean a system that can never be understood in a well-defined way. This interpretation, however, invites troublesome disputes over what is to count as well defined, and it also prejudges the question of whether human knowledge will ever be adequate for the system concerned. For instance, Schroedinger's equations are mathematically well defined, but they concern quantum phenomena that many would regard as a paradigm case of ill-definedness; and though the Copenhagen School believed this ill-definedness to be grounded at the ontological level, Einstein cited his conviction that "God does not play at dice," when interpreting quantum indeterminacy as a merely epistemological matter. Therefore, an ill-defined system is one with respect to which certain prima facie relevant types of theoretical description are inappropriate because they treat the system as being more well defined in a specific respect than it actually is. Minds (especially human minds) are ill defined in a number of ways; that is, there are several types of theoretical descriptions that one might expect to apply to mental phenomena but that are inappropriate because each wrongly assumes that minds are well defined in some specific way in which they are not.

Speaking now about our state of knowledge in respect to control, W. M. Wonham, from the University of Toronto, says that optimization of first-level, linear, multivariable systems is fairly well in hand. Nonlinear dynamics of lower levels can usually also be dealt with, at least numerically, and structural results are available here for specific classes of nonlinear systems.

Multilevel systems have been treated mainly from the viewpoint of optimization, using ideas of problem decomposition originating in mathematical programming. By contrast, the logic or synchronization problem of supervisory control that underlies the software engineering of real-time programs for concurrent control processes have scarcely received any attention from a theoretical viewpoint. Higher-level adaptive techniques in a logico-linguistic framework (inspired by artificial intelligence) have not yet made much progress as a domain of system control. It seems likely that the present decade will witness systems control, as a discipline, borrowing from and contributing to the theoretical foundations of computer science (i.e., logic and language conceived procedurally), just as, in the past, control science has been supported by Fourier analysis (1948-1957), differential equations and functional analysis (1958-1969), and abstract algebra (1968-1980).

All this leads to the commonplace observation that the modeler has to improve his tools. Such an improvement was suggested in

C. V. Negoita, 1979, *Management Applications of System Theory*, Basel and Boston: Birkhauser Verlag.

The message of this book is that system theory can guide us in deriving far-reaching conclusions from clearly stated premises, with the observation that "clear" does not mean "precise." The classical theory of systems is linked with the so-called qualitative analysis with linguistic values, with the conclusion that vagueness is not a liability. On the contrary, it makes for robustness.

To obtain a feeling for the flavor of linguistic models, the reader can refer to

F. Wenstop, 1980, Qualitative analysis with linguistic values, *Fuzzy Sets and Systems* **4**:99-115.

Wenstop discusses the use of quantitative models—where the attributes are variables—in the study of systems where human behavior on the individual level plays an essential role. The models specify how values of the variables are interrelated. The values represent quantities of the attributes in question.

Engineering and physics are examples of disciplines in which quantitative analysis has proved to be indispensable because of its deductive power. A quantitative model is deductive if there are consistent rules for deriving the output values when the input values are known. This is achieved automatically when the variables are numeric. The use of quantitative analysis in the soft sciences has, despite great effort, been less spectacular because of the difficulty in constructing reliable and valid measures. Most conceptualized attributes regarding human behavior, for instance, are not a priori operational. They cannot be directly measured. To make them operational, one must define them in terms of a measurement procedure. "Ability," for instance, is an attribute with a clear, although inprecise, meaning. It can intuitively have low or high values, but if the values are going to be numerical the attribute will have to be operationally defined. The meaning of the linguistic values such as "low" or "high" can be modeled with respect to a *psychological continuum* representing the range of the attribute on an interval scale. By this trick the meaning is represented as a table with two columns of numbers. Thus, one is much less likely to be wrong when one is using many numbers instead of one. The many numbers in the table are given the meaning of a unique name. We resorted to vagueness as a hedge.

W. Alston, 1967, Vagueness, in *Encyclopedia of Philosophy*, P. E. Edwards, (ed.), New York: Macmillan.

To say that a word is vague is to say that there are cases in which there are no definite answers to whether it applies to something. Thus, "middle-age" is vague, for it is not clear whether a person aged 40 or a person aged 59 is of middle-age. To say there there is no definite answer to the question, Is a person aged 40 middle-age? is not to say that we have not yet been able to find the answer because of insufficient information. The status of this question is quite different from that of the question, Is Mars inhabited? where we have a pretty good idea of what information, if obtained, would settle the matter. Our inability to say whether a 40-year-old man is of middle age is not the result of lack of information about such things as blood pressure and metabolic rate. No additional information would settle the matter, except indirectly by leading us to tighten up the meaning of the word. The indeterminancy is due to an aspect of the word rather than to the current state of our knowledge.

The word "vague" is commonly used quite loosely to cover a variety of features of discourse that should be distinguished. We should distinguish vagueness, as just defined, from lack of specificity. In both cases vagueness comes from the lack of precise boundaries between application and nonapplication along some dimension (age). This kind, which Alston calls "degree vagueness," furnishes the standard examples in most discussions of the subject, presumably because it is the easiest to discern and analyze; but it is not the only important kind. Another, more complex, source of indeterminancy of application is found in the way in which a word may have a number of logically independent conditions of application.

Any attempt to completely rid our language of vagueness is chimerical. For all practical purposes we must use words to provide more precise definitions of other words, and the task of removing the fresh elements of vagueness that come to light with each new definition seems to be endless. A practical attempt to model vagueness was made in 1965.

L. A. Zadeh, 1965, Fuzzy Sets, *Information and Control* **8**:338-353.

This historical paper suggested changing the definition of the characteristic function of a classical set. To each element in the set, Zadeh added a degree of membership in that set. This degree is a real number between 0 and 1. The characteristic function of a fuzzy set, instead of mapping to the set of two elements (a binary choice of being in or out of the set) is a mapping to a portion of the real line, allowing a continuum of possible choices. The definitions of basic operations on sets, union and intersection, were also extended.

Ideas that introduce a fundamental new way of viewing the world that captures the imagination of the scholarly community are rare indeed. Zadeh's paper suggesting fuzzy sets as models of vague concepts is certainly one. From that moment system theory and vagueness began to be considered together, at least at the level of a general theory. Such an attempt was made in

C. V. Negoita, and D. A. Ralescu, 1975, *Applications of Fuzzy Sets to Systems Analysis*, Basel: Birkhauser Verlag, and New York: Halsted Press.

This compact presentation of the algebraic theory of fuzzy sets is based on the so-called theorem of representation. Any fuzzy set is viewed as a family of crisp sets—its levels. Any fuzzy system is therefore a family of crisp systems. A rigorous mathematical approach can be tried, and the book goes along this line. This attempt led to enthusiasm, critique, and debate. For its history and prehistory, see the 1984 autumn issue of the journal *Human Systems Management*.

What is much more important is that the theory of fuzzy systems opened the door for a structural interpretation of purposeful systems, defined against a functional interpretation promoted in the 1940s by the pioneers of cybernetics. The functional interpretation started with

A. Rosenblueth, N. Wiener, and J. Bigelow, 1943, Behavior, purpose and teleology, *Philosophical Science* **10**:18-25.

This paper also captured the imagination of the scholarly world. The hope of glimpsing the secret of self-regulation and, therefore, of intelligent behavior may have driven many intellectuals in all fields to become cyberneticians. This interest led to the discovery of predecessors, going back to Plato.

M. Draganescu, C. Balaceanu, P. Golu, and A. Giuculescu, (eds.), 1981, *Odobleja between Ampere and Wiener*, Bucharest: Academy Press.

This contributed volume comments on the book *Psychologie Consonantiste* by Stefan Odobleja (Librairie Maloine, Paris, 1938). The interest for this book was triggered at the Third and Fourth International Congresses on Cybernetics and Systems held in Bucharest in 1975 and Amsterdam in 1978, where the author claimed the parenthood of a generalized cybernetics. The editors consider that he had a clear understanding of cybernetic processes, which he described without mathematical treatment or Shannon's technical notion of information, introduced in 1948. Their conclusion is that Odobleja can be ranked between Ampere, who used the word "cybernetics" in a classification of sciences, and Wiener, who formulated modern cybernetics and concluded that our capacity to control is inherently incomplete and necessitates a statistical approach. Manfred Kochen, in a paper published in *Human Systems Management* (1984, **4**:306–308), says that this approach was not sufficient for the analytic design of complex systems, a matter of dispute between Zadeh and R. Kalman that provided the initial impetus for the development of a fuzzy set theory. This seems to be true, because a structural approach to purposeful systems was developed in the framework of fuzzy systems. More details can be found in

C. V. Negoita, 1980, Pullback versus feedback, *Human Systems Management* **1**:71–76.

C. V. Negoita, 1981, *Fuzzy Systems*, Tunbridge Wells, England: Abacus Press.

C. V. Negoita, 1984. Pullback in knowledge engineering, *Human Systems Management* **4**:229–235.

C. V. Negoita, 1985, *Expert Systems and Fuzzy Systems*, Menlo Park, CA: Benjamin/Cummings.

A. Dinola, and A. Ventre, 1982, Ordering via fuzzy entropy, in *Fuzzy Information and Decision Processes*, M. Gupta and E. Sanchez (eds.), Amsterdam: North-Holland.

The desideratum for a new approach was explicitly formulated also by the fathers of artificial intelligence. An example is

M. Minski, 1984, Adaptive control: From feedback to debugging, in *Adaptive Control of Ill-Defined Systems*, O. Selfridge, E. Rissland, and M. Arbib, (eds.), New York: Plenum Press.

Minski says that there might be some basic cybernetic mechanism for acquiring knowledge and that somebody has to hit on a good way to describe control systems for computers.

Chapter 1

SIMULATION MODELS

COMPLEX SYSTEMS ENGINEERING

In the beginning, *engineering* was simply defined as the art of constructing and using machines. Today, it is known as a technical science, where *technique* is understood as a method of doing something expertly, and *science* means knowledge arranged in an orderly manner and obtained by observation and testing of facts. From the beginning, engineering drove science. Lately, the sciences have seemed to be independent processes of investigating physical, biological, or human phenomena. Later, a combination of analysis and design was aimed at first understanding how an existing system works and then preparing system modifications to change the system behavior.

The ideal to which much of the quantitative social and behavioral science literature strives is to develop "laws" of human behavior that rival in majesty and scope those laws for the behavior of natural systems bequeathed to us by physics. Some scientists argue that this envy of physics is completely misplaced, because no such laws exist, at least as that term is understood in natural sciences. They say that to qualify as a law of nature rather than as a simple empirical relationship among observables, a model must be

Independent of the particular physical situation in which it is observed: We cannot have one law of energy conservation for nuclear reactors and another for the kitchen stove.

Analytic: Local space-time information is enough to determine the model, and we need not account for what is happening in a distant location or time.

> *Invariant:* The model should not depend upon the scale or language used to describe it; that is, it should be coordinate free.

The last condition is the most interesting. Basically, invariance means that a given model remains the same under any relabeling of the observable. Technically, changes form *groups*, and in physics it is sufficient to verify the invariance conditions under some sufficiently large group. In the social and behavioral sciences the problem arises because there are no mathematically interesting groups that can be readily interpreted in terms understandable in the social or behavioral context. Some authors say that until such groups are identified the search for laws in the social sciences will remain a chimera. Others say that if we keep looking we could find them.

Anyway, the point of contact between the existence of laws and the concept of complexity should now be clear: If laws exist, then the invariance requirement asserts that there is only one description—that is, the system is *simple*. On the other hand, if the system has only a single description to which all others can be reduced, then the description is invariant and is a candidate for becoming a law.

Perhaps this is the reason why system engineering is dominated by a concern about complexity. This complexity is brought about because of the multitude of competitive perspectives that surround almost all contemporary issues. Because of complexity, there is no unique description, and the key to system engineering is the overlapping of many models. Our inability to make unique statements about complex behaviors is a fact we have to accept and adjust to. Complexity is associated with description rather than being thought of as an intrinsic property of natural processes. Hence, we may well consider reducing complexity not by changing the systems but by changing our views about it.

People who have attempted to analyze complex systems have frequently remarked that arithmomorphic models (mathematical models whose variables are numbers) are too poor to convey all the possible descriptions that a person can have in mind. For this reason social scientists require the medium of ordinary, natural language and its richer content.

Each word of a natural language is inherently vague—an arresting but not disturbing feature. Indeed, this is a very advantageous feature. Vagueness is not thought of as inherent in the natural world but in the use of words. Vagueness is used by humans to cope with complexity. Using vague words, humans get invariant models.

The word *cat* is

> *Invariant* because it does not depend on the number, size, or color of existing or future cats. The meaning is the same whether translated in Russian, French, or Swahili.

Analytic because local space-time information is enough to determine its meaning (we do not need to see all the cats in the world in order to have the meaning).

Independent of all particular cats that can be observed (we do not have a word for every cat in the world).

Therefore, we can use linguistic models to avoid complexity.

Today system engineering is linked with the *organization of knowledge.* Contemporary research efforts in system engineering place major emphasis on human-machine interactions and the use of natural language interfaces for effective management information systems. System engineering complements and enhances traditional engineering activities by emphasizing the information basis for new skill-based, rule-based, and formal reasoning-based assistance, rather than responding to the dictates of behaviorally insensitive and inflexible software. The idea of a linguistic model suplements the idea of simulation, and this is a step farther in engineering and science.

MODELS VERSUS PROGRAMS

In everyday language, to "simulate" means to pretend to be, to assume the mere appearance of something. In science and engineering, "simulation" is the forming an abstract model from a real situation in order to understand the impact of modifications and the effect of introducing various strategies.

The main objective of any simulation program is to aid the system analyst in projecting what will happen to a given physical situation under certain simplifying assumtions. It allows the user to experiment with real and proposed situations, which are otherwise impossible or impractical.

The major advantage of simulation is that it permits experimentation without modifying the real situations. The objective of any simulation program is to predict how a system will perform before it is built. Numerous alternative modifications to a situation may be included in a simulation, and their results can be comparatively studied systematically.

To simulate a connected series of actions, we must produce a model of the process. A model of an object or of a process could be another object or process or a description of them (mathematical or linguistic). It is possible to construct a physical model whose behavior represents the system being studied. The best-known examples of physical models are scale models used in wind tunnels and water tanks to study the design of aircraft or ships. Accurate deductions about the performance of a full-sized system can be made from the scale model.

The analogy between the process and its physical model is established when they both obey the same physical law. A real process and its mathematical description are models of each other, and the performance of either can be studied with the other. In practice it is simpler to modify the mathematical description than to change the real process.

There is no unique model, since the form the model takes depends on the information selected from the process. The model is finite and therefore cannot produce information on every aspect of the original process.

Consider a process of population growth, the problem of squirrel reproduction based on the following axioms:

We start with one squirrel in year 1.
A given squirrel reproduces first at age 2 and then every year thereafter.
A reproduction year for a squirrel involves exactly one new birth.
All squirrels live forever.

According to these assumptions, the number of squirrels S_k in year k can be thought of as generated by the difference equation

$$S_{k+1} - S_k = S_{k-1}$$

which says that the number of last year's squirrels, S_{k-1}, is the difference between next year's squirrels, S_{k+1}, and this year's squirrels, S_k. This model can also be put in the form of a recurrence relation:

$$S_{k+1} = S_k + S_{k-1}$$

In a more suggestive form, we write

$$\text{Future} = \text{Present} + \text{Past}$$

Let us use a computer to print out the number of squirrels per year that is less than a predetermined number. The computer, is not intelligent. It cannot analyze a problem and come up with a solution. The programmer must arrive at the solution and communicate it to the computer. But using a computer involves more than simply writing a program. A programmer must understand and analyze the problem. He or she develops a general solution called an algorithm, a step-by-step description of logical sequences of actions for solving a problem in a finite amount of time.

The task of deriving a program may be divided broadly into two subtasks: establishing the model structure and supplying the data. They are usually so intimately related that neither can be done without the other. Assumptions about the system direct the gathering of data, and analysis of the data confirms or refutes the assumption. Therefore the following scheme has to be observed:

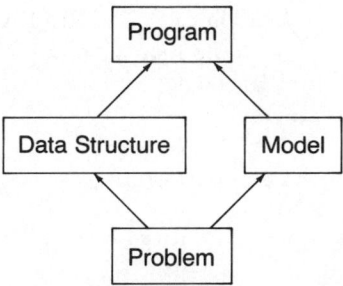

We use programs to get the computer to work for us. Programs are written in a special programming language, which is a set of symbols, and special words. When designing an algorithm, a programmer should keep in mind the things a computer can do according to a specified programming language.

If we want a PASCAL program for our algorithm, then we have to specify a heading, declarations, and statements. The heading is composed of the reserved word PROGRAM, followed by a user-defined name for the program, followed by a list of the files used by the program.

In our case, the heading will be

```
PROGRAM SQUIRRELS (OUTPUT);
```

Next in the program the declaration section gives information to the compiler to use in the translation phase. The program operates on data. In our case, all the values are numbers, and all the data may be of the simple type INTEGER.

To express the algorithm in a computer program requires two additional steps. We have to let the program know what we know, and we have to let it tell us what it knows. What we know is the given number of years or a given number when the calculation should stop. We will give this information to the program by setting a constant equal to this value.

The VAR section causes the variable names PAST, FUTURE, PRESENT to be assigned to three memory locations, but no values are put into these places. The places are needed to hold the results of the calculations.

The executable statement of the program is delimited by the pair BEGIN-END. The statements between BEGIN and END are translated into the machine language that the computer can execute. After the translated version of the program is in memory, control is turned over to the first statement after BEGIN. In other words, what the program says to do is actually done. Changing the value of a variable is done through an assignment statement.

To write the values representing the number of squirrels per year, we shall use a loop, a program section that executes repetitively, changing certain internal variables at each execution pass. The REPEAT-UNTIL statement is a looping control structure with the loop condition tested at the end of the loop after the reserved word UNTIL.

Here is the complete program

```
PROGRAM SQUIRRELS (OUTPUT);
CONST MAXS = 10000;
VAR   PAST,PRESENT,FUTURE:INTEGER;
BEGIN
   PAST:=0;
   PRESENT:=1;
   REPEAT
        FUTURE;=PAST + PRESENT;
        WRITE  (FUTURE);
        PAST:=PRESENT;
        PRESENT:=FUTURE;
   UNTIL PRESENT >=MAXS;
END
```

In this program the model is the line

```
FUTURE:=PAST + PRESENT
```

The rest of the program is *overhead*; that is, it takes care of variables.

There is a trade-off between keeping a program simple, to reduce cost, yet detailed enough to be resonably accurate. Assumptions made to simulate some process tend to be a key to the accuracy of the results. The more the assumption is simplified, the less reliable are the results; whereas the more details that are included, the better (generally) are the results. No matter how detailed the program, the prediction obtained will be inaccurate if the initial assumptions are incorrect or if the situation studied changes over time. In either case the simulation program becomes invalid as a prediction tool. Clearly the model is more important than anything else. In the sequel we will concentrate on its realistic formulation.

REALISTIC MODELS COMPLICATE THE PROGRAMS

So far, we described system simulation as the technique of solving problems by following the changes over time of a dynamic model. We will turn now toward system analysis, aiming to understand how an existing system or a proposed system operates. Usually, the behavior of the system is known, but the processes that produce that behavior are not. Therefore, hypotheses are made on a likely set of entities that can explain the behavior. This is what we did when we tried to describe the population growth. Unfortunately, our model representing the squirrel population growth has a behavior that does not match the behavior of a real process: According to our model, the number of squirrels will grow boundlessly, which is not realistic. Definitely, we have to formulate a refined set of assumptions. First, it is not hard to observe that the squirrels do not live forever. Instead of

$$\text{Future} = \text{Past} + \text{Present}$$

we have to consider something like

$$\text{Future} = \text{Past} + \text{Present(born)} - \text{Present(lost)}$$

We have now four variables: FUTURE, PAST, PRESENT.BORN, PRESENT.LOST. Clearly, making the model more realistic, we increased the complexity of the program. We now need more space in the memory and more time for processing the model.

But this is not the only possible improvement. We cannot start with only one squirrel in year 1. There are many squirrels at the beginning of the simulation process, but nobody, absolutely nobody, can count them to tell us with minute precision how many squirrels do exist on the earth right now. All we know is that there are *many* squirrels, and, in the best case, an expert can approximate their number with an evaluation, giving us the meaning of or the context of the linguistic label MANY. But we have seen that our program needs precise data on which to operate—integer or real numbers, nothing else. This assumption is very restrictive and, as we have seen, one we do not always encounter.

Sometimes we need linguistic variables to eliminate additional minute specifications. For instance, in our model it is not certain that the squirrel reproduces every year. We do not know the precise and extremely complicated law, if any, governing the reproduction process, but we are pretty sure that it is absurd to say that a reproduction year for a squirrel involves exactly one new birth. Our common sense forces new assumptions. We are inclined to say

We start with many squirrels.
A reproduction year involves a few births.

If so, then with these two prudent statements we achieved *robustness*.

Confidence in the truth of a vague assertion is justified precisely because of its vagueness, which makes it compatible with a whole range of observed facts. The law of population growth can acquire the fixed and absolute certainty of common sense only by sacrificing the minuteness of detail. A vague formulation is *structurally stable*—that is, unique and covering a large family of particular situations. A structurally stable system is a system whose behavior is not drastically altered by a slight change in its structure. Many and few can be evaluated differently, according to different contexts, but the model remains the same. We say that vagueness, far from being a liability, is a blessing. This statement deserves an explanation. Building a model is based on understanding. The event of understanding is one in which a *person* opens up and thereby extends or broadens the image that describes the tacit perimeter of his or her subjective view. Image is the range of vision that includes everything that can be seen from a particular point. The notion of blurring is relevant to this placement. A coarsely sampled and quantized portrait could be recognized more easily if viewed at a distance. This has been put into practice by painters for a long time. They *pull back* in order to understand the whole. "Many" and "few" are words expressing quantities. We pulled back into vague assumptions to get structural stability. But with this type of data we cannot run our program. PASCAL and any other programming language do not accept this type of data. They accept only numbers. Therefore, if we want to do the simulation we have to go back to numbers. But how?

One possibility is to model the label "many" with a number, say 100, just because we have the feeling that 100 squirrels can approximate the meaning of many squirrels. But 99 squirrels do not stop being many because one squirrel has been subtracted from the pack. Therefore a pack with 99 squirrels still represents many squirrels. If we continue arguing in the same pattern, subtracting yet another squirrel from the pack does not change its status. In this way we can reach the absurd conclusion that a pack with one squirrel is a good representation of many. There appears to be no good reason for stopping at one number rather than another. It is difficult to see why the chain of argument should ever be broken, but it is ridiculous to conclude that a pack of two squirrels is a good representation for many. Almost no one considers a two-squirrel pack many. Almost everyone considers a two-squirrel pack few. In between there is a gradual variation. Anybody confronted with this representation problem can see that the puzzle arises from deficiencies in modeling the concept many by a unique number. It is a feature of our use of a vague concept, such as many, that there is no determinate *point* at which a transition from a clear case to a borderline case occurs. The occurrence of vague expressions thus testifies at once to the existence of experienced continua of qualities and also to the absence of fixed habits of discrimination between segments of such continua.

A vague concept whose nature implies the lack of a criterion for deciding its truth or falsity shows the inapplicability of the principle of the excluded middle. In other words, vague concepts cannot be handled according to classical, two-valued logic.

Moving along the string of integers or real numbers, we observe its elements. A two-valued logic is quite sufficient to assess the existence of these elements. Handling vague labels, we feel the need of a multivalued logic or, better, of an infinite-valued logic to assess the degrees of existence mirroring the degrees of membership of different numbers in our label. Thus a predicate need not be simply true or false, but it may be partly true to any degree. Let us assume that this degree is a real number in the interval [0, 1]. Consider the label "many." Different numbers have different degrees of membership with respect to this label. For example, 100 might have degree 0.9, 1000 might have degree 1.0, 10 might have degree 0.1, and numbers in between might have intermediate degrees. Different individuals will have differing opinions as to whether a given number should belong to or should be described as many. A possible representation could be

Many

1	0.1
2	0.5
3	0.8
4	1.0

This is a function

```
Many:Integers → [0,1]
```

with the properties

```
Many(1)=0.1    Many(4)=1.0
```

Note that in this representation we have imposed a continuity in the membership distribution. In other words, the integers with the highest membership are clustered around a given real interval. This fact allows us to easily understand the semantics of the table. Another requirement imposed was that of *normality*: Among the points of the right column with the highest membership value, there exists at least one that is completely compatible with the predicate associated with the table; that is, it takes the value 1. With these conditions we say that our table is a *fuzzy number*.

Our model is the same, but the values taken by the linguistic variables are tables. A table with rows and columns is a two-dimensional array. Fortunately, an array is a built-in data structure of PASCAL, and the linguistic variable MANY can be declared

```
VAR MANY: ARRAY[1..2] OF ARRAY[1..4] OF REAL
```

We refer to an element of this array by its name and its index. For instance, the first element of MANY would be referenced MANY[1][1].

The addition of two tables requires that the base on which the fuzzy numbers are defined to enlarge and to shrink. For example, if two fuzzy numbers defined over [1, 2, 3, 4] are added together, the resulting fuzzy number is defined over [1, 2, 3, 4, 5, 6, 7, 8], according to the following possible combinations:

$$2 = 1 + 1 \quad 3 = 2 + 1 \quad 4 = 3 + 1 \quad 5 = 4 + 1$$
$$3 = 1 + 2 \quad 4 = 2 + 2 \quad 5 = 3 + 2 \quad 6 = 4 + 2$$
$$4 = 1 + 3 \quad 5 = 2 + 3 \quad 6 = 3 + 3 \quad 7 = 4 + 3$$
$$5 = 1 + 4 \quad 6 = 2 + 4 \quad 7 = 3 + 4 \quad 8 = 4 + 4$$

The result of adding Many and Few, defined as

Many

1	0.1
2	0.5
3	0.8
4	1.0

+

Few

1	0.8
2	1.0
3	0.8
4	0.5

is the new table

Many + Few

1	0.0
2	0.1
3	0.5
4	0.8
5	0.8
6	1.0
7	0.8
8	0.5

where the degrees are obtained according to the rules of fuzzy arithmetic to be given later.

In most computer languages, as in PASCAL, arrays cannot have variable dimensions. The inefficiency of inserting and deleting elements from a list sequentially allocated in an array occurs because the order of the elements is recorded implicitly. Adjacent records of the list must be in adjacent memory locations, so more than one record may have to be moved for an insertion or a deletion. This movement can be avoided if the order of the elements is recorded explicitly. In particular, a linked-list implementation associates with each list element a pointer to indicate the address at which the next element is stored. Each element of the linked list will contain a number from the first column of the table, a number from the second column of the table, and a pointer to the next element.

The implementation of the addition operation will take full advantage of the flexibility of this linked-list representation—that is, the abilities to insert new elements at any point easily and to expand and shrink the list size dynamically at execution time. Basically, the implementation of Many + Few is as follows: For each element in the list Many, one will step through the elements of the list Few; and for each element of the list Few, one will (potentially) construct a new element for the list Many + Few, which will be inserted in its proper position.

THE THEOREM OF REPRESENTATION SIMPLIFIES THE PROGRAMS

Consider the label Many, whose meaning is given in the table

Many

1	0.1
2	0.5
3	0.8
4	1.0

where, for instance, 3 belongs to Many with the membership degree 0.8. From this table we can extract *level sets* (i.e., crisp sets formed with elements from the *support* [1, 2, 3, 4]), namely those elements with a membership value equal to or greater than 0, 1, 0.5, 0.8, and 1.0.

For level 1.0 the level set is [4]. For level 0.8 the level set is [4, 3]. For level 0.5 the level set is [4, 3, 2]. For level 0.1 the level set is [4, 3, 2, 1]. All these level sets are ordered and form a chain; that is, they are included in each other:

$$[4] \subset [4, 3] \subset [4, 3, 2] \subset [4, 3, 2, 1]$$

According to the theorem of representation, any subjective evaluation defined on any support is equivalent to a family of intervals.

It is easy to see that the level sets form a set-through-time indexed by the membership degree. In this way the elements 4, 3, 2, 1 of the support have different degrees of existence in the fuzzy number Many. The element 4 is more existent than the element 3, and 3 is more existent than 2. Therefore we can represent the table Many by the new table

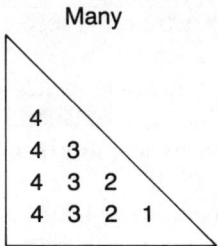

emphasizing the importance of the elements or their existence. Because column 4 is higher than column 3, the degree of existence of the element 4 is said to be greater than the degree of existence of the element 3.

The flexibility introduced by the linguistic approach is a valuable tool only if it becomes efficient. To achieve efficiency, we have to simplify the numerical representation. Because the family of level sets forms a chain, we can use only the first and the last level sets in the family. Therefore, the meaning of Many can be represented by the table

Many

[4]
[4...1]

This simplification is possible if in the definition of a fuzzy number we impose normality and convexity.

In general, any normal convex fuzzy number may be characterized by a 4-tuple (a, b, §, ¢), where [a, b] is the closed interval on which the membership function is equal to 1.0, § is the left bandwidth [(a − §), a], and ¢ is the right bandwidth [b, (b + ¢)].

Any crisp number can be represented in this form by (a, a, 0, 0), and our fuzzy number Many becomes (4, 4, 4, 0). In this way, if

$$\text{Many} = (a, b, \S, ¢) \quad \text{and} \quad \text{Few} = (c, d, @, \#)$$

then

$$\text{Many} + \text{Few} = (a + c, b + d, \S + @, ¢ + \#)$$

and we are back to simpler data structures, avoiding linked lists and complicated loops in our programs.

MODELS IN RELATIONAL DATA BASES

System simulation is a numerical technique using dynamic mathematical models. A dynamic model is essentially a set of equations whose variables can be grouped as *exogenous* (or input) and *endogenous* (or state).

"Endogenous" describes activities occurring within the system; "exogenous" describes activities in the environment that affect the system. A system for which there is no exogenous activity is said to be closed, in contrast to an open system, which does have external influences.

Dynamic models can be used to combine in a strictly mathematical manner an analysis of causes and effects with or without time lags and feedback relationships. A fundamental postulate is that the system or process to be modeled can be described by a set of state variables that are functions of time. The state of the system at time k is identified with the vector $x_1(k)$, $x_2(k), \ldots, x_n(k)$. The set of all possible states of a dynamic model is called the *state space*. Normally, a system can be controlled from outside, and the control can be characterized by a set of input variables, $u_1(k), u_2(k), \ldots, u_m(k)$. The set of all possible input vectors is called the *input space*. Any change in the state vector can be defined as a function **F** of the previous state and input vectors:

$$x(k + 1) = \mathbf{F}(x(k), u(k))$$

$$\Downarrow$$

The future state *Is* a function of present state and present input

The state equation describes a flow at various moments, and the function shows the connections between the components of state and input vectors. These connections reflect an existing pattern of organization. We say that an observer sees the system through a set of observables linked by the state equation. The greater the number of nonequivalent ways the modeler can see the system, the greater the number of possible counterintuitive models.

A simulation of a dynamic model during a time interval T consists of supplying initial values for the state variables and the input variables during T and then using the model equations to compute the values of the state variables for a number of times during T.

So far we have considered the case where x and u can be integers or fuzzy integers and where the state equation (the model) is an instruction (or a set of instructions) in a program that can calculate a trajectory

$$x(T) = H(x(0), u(0), \ldots, u(t - 1))$$

which describes the evolution of the system as a function H of the initial state $x(0)$ and the sequence of inputs.

This general picture applies to practically all known dynamical systems, provided that they are directed by a local determinism. However, this model gives rise to serious difficulties. Few phenomena allow a mathematical description—that is, for which the function $\mathbf{F}$ is known exactly and given by explicit formulas. Most real situations do not admit mathematical descriptions at all. On the other hand, in very complex fields, like economics, a multitude of mathematical models exist, although the implementation of any is practically nonexistent. Apparently, most managers have decided to use their heuristics—that is, methods of solving problems by inductive reasoning, evaluating past experiences, and moving by trial and error. Available empirical evidence seems to indicate that managers judgmental models do remarkably well, considering the complex nature of the economic problems. Managers using simple heuristic decision rules may be able to closely approximate the results of much more powerful optimizing techniques for aggregate planning and scheduling problems. Linguistic variables are used to transform reasonable rules of thumb into an operational model (see Chap. 2).

For a quantitative model the relationships between the variables would be expressed as equations. For qualitative models the relationships between linguistic variables are expressed as fuzzy conditional statements. Such a statement could be

IF $x(k)$ is High and $u(k)$ is Low THEN $x(k + 1)$ is High

where High and Low are fuzzy numbers. To efficiently implement this model, we must parametrize the variables. This step describes the universes over which the fuzzy numbers are defined. Two parameters are required for each variable in the model: the lower bound and the upper bound that the variable is expected to take on. An example could be:

Fuzzy number	Interpretation	Mnemonic
(0,0,0,2)	Very Low	VL
(0,1,0,2)	Low	L
(2,2,2,2)	More or less Low	MLL
(5,5,2,2)	Average	A
(8,8,2,2)	More or less High	MLH
(9,1,2,0)	High	H
(10,10,2,0)	Very High	VH

According to these notations, the following relational data base can be formed

Rule	IF x(k)	and u(k)	THEN x(k + 1)
1	L	L	MLL
2	H	L	MLH
...	...	...	...
n	VH	L	VH

Consider the single rule IF A THEN B, which is a relation between a premise A and a conclusion B. The degree of existence of this relation cannot be greater than the degrees of existence in each component; therefore, if we consider the function f describing this fact, we set

$$f_{A \to B}(x, y) = \min[f_A(x), f_B(y)]$$

Suppose now that a new premise A' is defined on the same universe. Then we have a relation

$$A' \to A$$
$$f_{A' \to A} = \min[f_A(x), f_{A'}(x)]$$

Combining these two relations, we have the diagram

$$\begin{array}{c} A' \\ \downarrow \\ A \to B \end{array}$$

By composition

$$f_{A' \to B}(x, y) = \min[f_A(x), f_B(y), f_{A'}(x)]$$

This will lead to

$$f_{B'}(y) = \max_x \min[f_A(x), f_B(y), f_{A'}(x)]$$

meaning that the inference is the closing of the diagram

$$\begin{array}{ccc} A' & \to & B' \\ \downarrow & & \downarrow \\ A & \to & B \end{array}$$

Now consider a strategy consisting of two rules. When this strategy is executed, the contribution from the individual rules are collected from the diagrams

$$\begin{array}{ccc} A'_1 & \to & B'_1 \\ \downarrow & & \downarrow \\ A_1 & \to & B_1 \end{array} \qquad \begin{array}{ccc} A'_2 & \to & B'_2 \\ \downarrow & & \downarrow \\ A_2 & \to & B_2 \end{array}$$

READINGS

L. Padulo, and M. Arbib, 1974, *System Theory: A Unified State Space Approach to Continuous and Discrete Systems*, Philadelphia: Saunders.

This is an encyclopedic treatment of the entire subject. Just about everything one might need to know about state-determined systems can be found here.

A system is state determined if, given its state at some time t_0 and the input of the system from time t_0 to some subsequent time t_s, we may specify what the new state will be at time t_s. Since $x(t_s)$, the new updated state at time t_s, is determined by t_0, $x(t_0)$, and the input function u over the time interval $[t_0, t_s]$, we express this functional dependence by

$$x(t_s) = g(t_s, t_0, x(t_0), u)$$

The function g is called the state transition map. To show that in the quadruple the first two entries come from the set T, the third entry comes from the set X, and the fourth entry comes from the set of admissible input functions U, one uses the notation of set theory

$$g: T \times T \times X \times U \rightarrow X$$

If at a time k the state is $x(k)$ and the input is $u(k)$, then the state at time $x(k + 1)$ will be $g(k + 1, k, x(k), u)$. This expression depends only on the value of u at time k. The value of u at time $k + 1$ arises simultaneously with the system settling into the state $x(k + 1)$ and cannot affect that state, but it can affect later states. Thus, we may replace g by some simpler expression $d(x(k), u(k))$, where we have made explicit the fact that the transition from state $x(k)$ at time k to state $x(k + 1)$ at time $k + 1$ depends on the input only through the value u takes at time k. In other words, we have replaced the specification of the complicated dynamics function $g: T \times T \times X \times U \rightarrow X$ (which is defined for arbitrary pairs of times and arbitrarily admissible input functions) by the much simpler function

$$d: X \times U \rightarrow X: (x, u) \mapsto d(x, u); \qquad x \in X, u \in U$$

The function d specifies a next state transition; that is, it describes the state dynamics in the locality of time k.

Linear system theory studies the case in which U and X are vector spaces and the map d is linear:

$$d(x, u) = Ax + Bu$$

Automata theory usually considers U and X to be finite sets but allows d to be nonlinear.

We may also imagine systems in which we cannot determine exactly what the subsequent states will be, no matter how fully we specify the state and no matter how carefully we specify the inputs. These systems are nondeterministic. For such systems the transition map can be of the form $d: X \times U \to 2^X$, where $2 = \{0, 1\}$. We interpret $d(x, u) \in X$ to be the set of probable successors to x when acted upon u.

A stochastic system has dynamics $d: X \times U \to XP$, where XP is the set of probability distributions on X. Often, a stochastic system may be approximately described by a deterministic structure that is stochastically disturbed.

C. V. Negoita, 1979, *Management Applications of System Theory*, Basel and Boston: Birkhauser Verlag.

Industrial and economic systems are much more complicated than the usual systems of physiscs. Borrowing the language of control engineering is one thing; applying its methods to management policies is another. This book attempts to select what seems to be important for management science from the burgeoning body of modern mathematical system theory.

This book is a revision of the lecture notes for an introductory course in system theory given between 1971 and 1976 at the Faculty of Economic Cybernetics in Bucharest, Romania. It is consistently characterized by the experience of the author as head of the research laboratory at the Institute of Management and Informatics there, and it owes much of its form to the comments and contributions of the participants in seminars at that institute.

Such a contribution was C. V. Negoita, D. Ralescu, and T. Ratiu, 1977, Relations on monoids and realization theory for dynamic systems, in *Modern Trends in Cybernetics and Systems* (Proceedings of the Third International Congress of Cybernetics and Systems, Bucharest, August 1975), Berlin: Springer-Verlag.

This article tries to give a new approach for realization theory—the theory of building a structure from a behavior. The special feature of this approach is the equivalence between the category of reachable systems (with a given behavior) and the category of right congruences on a monoid. This approach enlightens the role played by the Nerode equivalence in minimal realization theory and shows that the minimal realization is isomorphic to a direct limit of realizations. If the category of right congruences is replaced by the category of two-sided congruences, this is proved to be equivalent to the category of reachable systems with a given behavior, whose state space is a monoid and whose reachability map is a monoid homomorphism. The Myhill realization is then the final object of this category, viewed as a direct limit. These results can be applied to a class of latticeal systems and, particularly, to the class of fuzzy systems.

The problem of realization is important when speaking about control synthesis, that is, the design of a control law for a system known from its behavior. The control problem is admirably presented in the following outstanding book.

W. M. Wohnam, 1979, *Linear Multivariable Control: A Geometric Approach*, 2nd ed., Lecture Notes in Economics and Mathematical Systems, Vol. 101, Berlin: Springer-Verlag.

This book introduces a detailed treatment of the so-called internal model principle: A control system for which both internal stability and output regulation are structurally table properties must utilize feedback of the regulated variable and incorporate in the feedback loop a suitable model of the dynamic structure of the exogenous variables that the control system is required to process. In 1976 (*IEEE Trans. Systems, Man, Cybernetics* **SMC-6:**135-140) Wohnam launched the idea of an abstract internal model principle and obtained a theorem that ensures, under some conditions, that this principle works even for systems with state spaces represented by ordinary sets and state transition maps represented by ordinary functions. Giving up the rich structure offered by normed linear spaces and linear maps, he obtained a more general result, but, at the same time, it appeared that the requirements of integral stability and structural stability must be given up, since they make sense at least in topological spaces. The property of structural stability was restored when the internal model principle was extended to fuzzy systems. This restoration was made in

C. V. Negoita, and M. Kelemen, 1977, On the internal model principle, in *Proceedings of the 1977 Conference on Decision and Control, New Orleans*.

Evaluating the spaces, the internal model principle remains valid. A theorem was proved according to which the evaluations in the controller must reflect the evaluations in the exosystem. To illustrate these ideas, imagine an orchestra conductor (the controller) who feels the preferences of the composer (the exosystem).

G. Gordon, 1969, *System Simulation*, Englewood Cliffs, NJ: Prentice-Hall.

This is a classical book centered on computer applications. The author, a member of the scientific staff of the IBM New York Scientific Center, created GPSS (General Purpose Simulation System), a widely used simulation language. The book offers a broad and highly functional treatment of simulation and its applications to engineering, the sciences, and management, and is not limited to a single technique or computer language. Instead, it introduces simulation principles through hand-computed examples, demonstrates their translation into FORTRAN, and includes solutions worked in five different simulation languages, including 1130/CSMP, 360/CSMP, DYNAMO, SIMSCRIPT, and GPSS.

The book offers examples of applications in engineering and biological systems, industrial and economic systems (industrial dynamics), switching systems, and inventory control. It provides lucid examples and full programming information for solving problems in different simulation languages. It discusses techniques applicable to both continuous and discrete systems and compares their relative merits. It explains the probability theory and statistical techniques involved in constructing valid models and in analyzing results. It seems to say everything. Yet the experience of the last 20 years has shown that there are many other things to say.

V. Vemuri, 1978, *Modeling of Complex Systems*, New York: Academic Press.

As the author states, scientific and technical developments in the past few decades have set the stage for an era characterized by bigness, by systems that are more complex than the familiar engineering systems on which so many textbooks are based. We are now speaking of large systems characterized not only by their geometrical largeness but also by a structure behavioral or social in its nature. The term "large scale" is a subject of value judgments. As it stands today, the theory of large-scale or complex systems, if such thing exists, is more a state of mind than any specific amalgam of methods or philosphies. The author feels, however, that sound mathematical and logical thinking must occupy an important position in any large-scale systems theory, to this end, bits of knowledge have been collected and organized in this book to fill some of the needs outlined earlier.

Although, almost inevitably, there are one or two areas in which the text barely goes beyond exhortations to good housekeeping, the greater part of it deals with practical, proven methods of system analysis. The target audience is defined as the new breed of students interested in a multidisciplinary approach to complex problems of contemporary societal interest. On the assumption that current societal problems will inevitably catch great numbers of these students unprepared, Vemuri's approach is to focus on appropriate means of analysis. The author knows that complex systems are not static, that many large-scale system problems are characterized by a conflict of behaviors, and that a significant practical aspect of a problem is one of learning enough about the system to permit the development of a meaningful policy of operation.

Examining how to handle complex systems, Vemuri observes that years of experience clearly show that there are four basic approaches to the scientific truth:

1. *The Leibnitzian approach*, based on the premise that truth is analytic. Therefore, a system can be defined completely by a formal or symbolic procedure.
2. *The Lockean approach*, based on the assumption that truth is experimental; this implies that the validity of a model does not rest on any other assumption. Among the Lockean methods of approaching complex systems problems, special mention is made about the Delphi technique, where opinions of a large group are required to treat an issue adequately.
3. *The Kantian approach*, based on the assumption that truth is synthetic. That is, experimental data and a theoretical base are inseparable. This approach can be incorporated into the Delphi technique, ideally suitable for ill-structured problems.
4. *The Hegelian approach*, based on the assumption that truth is conflictual. The union of conflictual images leads to a more adequate grasp of the nature of things until finally all possible points of view, with all their seeming conflicts, become the constituents of one comprehensive system. One important fact underlying the dialectical approach—and this is what we are talking about—is the recognition that data is not information and that information results from an interpretation of data.

J. Sutherland, 1978, *Societal Systems: Methodology, Modeling, and Management*, New York: North-Holland.

In this book a system is viewed as a collection of individuals bound together by sets of interests and pursuing a set of ambitions. The author is partially in sympathy with those who suggest that we might do well not trying to manipulate societal systems, but he suspects that humans will continue to do so. He recommends an explicitly *transdisciplinary* mode of inquiry, emphasizing that complex models must use synthetic constructs as vehicles for countering social science parochialisms.

A *synthetic construct* is a model that seeks to reconcile competitive paradigms by raising the level of inquiry to a higher level of abstraction. Sutherland argues that by using synthetic constructs one can find causal relationships between attributes that are not apparent from the perspective of any particular discipline. He sees the dialectical engine as a mediating tool, serving to connect the individual to his cultural context. The word "dialectical" connotes a situation of conflict. Specific behaviors are products of a resolution of competitive predicates. It is through the office of synthesis that images become mixed. Sutherland sees all complex phenomena as collisions of separate and competitive interests. According to him, the intrinsic bases of behavior are the properties of the individual, the cultural predicates are the creative individuals, and the contextual variables are the attributes of reality.

Sutherland's clear premise is that culture exercises a tremendous influence on individual behavior. Most complex systems, he says, have been analyzed only in one dimension. Trying to understand numerical simulations, he introduced the "corridor concept." The danger of analyzing a system "along a single corridor" can be avoided by linguistic descriptions.

S. Tyler, 1978, *The Said and the Unsaid*, New York: Academic Press.

There is now evidence of a growing awareness of scientists that mathematical theories, if they claim to deal with complex phenomena, must be subjected to the same kind of critical discussion that is widespread in the natural sciences. The extent to which a theory constitutes a valuable addition to our knowledge depends very strongly on its ability to be challenged and to successfully withstand the challenges. It follows that criticism should be an integral part of any scientific activity and that efforts to challenge existing theories are necessary for scientific progress.

Tyler challenges the intellectual poverty of formalism. He says that *language* is not only for representing ideas but is equally a means for expressing beliefs. His focus shifts from the language itself to what people do with language.

Meaning, he says, is a matter of interpretation rather than the automatic translation of preordained instructions. Meaning is therefore a variable.

Functionalism, by emphasizing what we might call the outer appearances of language, asks not how the words are interpreted but how they create appropriate effects in others. Clearly, conventionality is not possible if each of us has a different experience of the world. In fact, the experience of each of us must be both the same and different. Tyler notes that formalism errs on the side of sameness, and functionalism errs on the side of difference.

Where formalism seeks to exorcise vagueness, the functionalist encourages us to look at the uses of vagueness. Not a problem to be overcome, vagueness is a necessary feature of language, which we suit to our own ends. The formalist's dream of explicitness contradicts our common sense. Without vagueness, communication becomes simply impossible.

As with vagueness, so too with silence. The formalist forgets that what is not said, either by way of implicitness or through silence itself, is often more important than what is said. Silence may communicate what is beneath words or beyond them, but in either event is part of our communication.

C. V. Negoita, 1981, *Fuzzy Systems*, Tunbridge Wells, England: Abacus Press.

Trying to apply mathematics to soft sciences, researchers are susceptible to trends of fashion. The systems approach has shown their readiness to accept a new methodology verified only in physics. There can be little argument that the expertise of system theory is potentially relevant to soft sciences. But the way in which this potential can be fruitfully realized is less obvious.

First, conventional systems approaches have tended to assume that what needs to be done is to survey the whole system with one model, no matter how big. But a major problem in complex systems is the existence of a large number of models based on different theories. In Greek, *theoria* means view. One way to define a consensus among equally creditable conflicting models is to integrate these views. The modern philosophy of complex systems assumes that a comprehensive enough family of models will allow us to reconstruct the original system.

Second, recent developments in global simulation models and the application of control theory in system engineering have brought into question the assumption that quantitative models can represent complex systems.

It is now clear that a successful system approach depends critically on the potential of the modeling method. Some form of modeling paradigm is necessary for the advance of the subject. To use simulation models in dealing with large-scale, highly complex systems, like those concerned with power distribution, transportation, and economic modeling, which do not lend themselves to precise analysis, we need a change in the methodology.

To grasp the whole, the modeler has to *pull back* and reduce the world to manageable proportions. An important part of this procedure involves the establishing of a perspective. The perfect expression of this procedure is a metaphor employed by Tocqueville in 1835:

> My present object is to embrace the whole from one point of view: the remarks I shall make will be less detailed, but they will be more sure. I shall perceive each object less distinctly, but I shall describe the principal facts with more certainty. A traveler who has just left a vast city climbs the neighboring hills; as he goes further off, he loses sight of the men whom he has just quitted; their dwellings are confused in a dense mass; he can no longer distinguish the public squares and can scarcely trace out the great thoroughfares; but his eye has less difficulty in following the boundaries of the city, and for the first time he sees the shape of the whole. . . . The details of the immense picture are lost in the shade, but I conceive a clear idea of the entire subject.

The paradox of wholeness and distortion is not considered any mere accidental or marginal phenomenon. It is placed now at the center of scientific thinking. Outside the paradox we are not able to understand the world. We have to learn how to identify the paradox in its diverse manifestations. If this paradox has up to now been considered as a symptom of a pathological state, in recent years it has appeared more frequently to indicate normality.

A thesis advanced in this book is that system theory is currently reaching the responsive frontier where the new skills of handling vagueness become critical.

Chapter 2

LINGUISTIC STRATEGIES

This chapter is devoted to rule-based techniques for control of processes. Such techniques emphasize an alternative approach to developing a program. The traditional approach relies on analyzing the system to be controlled or the process about which decisions have to be made. The motivation for such an approach is that once a quantitative model of the controlled system is available, then all decisions regarding it can be made via an algorithm. The rule-based approach shifts attention from the controlled system to an existing decision maker, who articulates the strategy and expresses it linguistically as IF-THEN rules.

THE ROTARY CEMENT KILN CONTROL

Everyone knows that gray powder, after being wet, becomes hard like stone and is used for building, but not everyone knows the technology of producing this powder. Cement consists of those hydraulic limes that contain silica and therefore set quickly. It is manufactured by grinding clinker, a mass of rough, hard, slaglike material left in a kiln after coal has been burned. The clinkers are produced in a kiln by heating a mixture of limestone and clay. If the kiln process is wet, the raw material mixture is prepared as a slurry and is fed to the back of the kiln, which is a long tube of steel slightly inclined and rotating. Through rotation, the material is slowly transported and heated in a counter current with very hot gases. These gases are pulled through the kiln by a fan and controlled by a damper.

Principal measuring points are the O and CO_2 percentages in gases and the kiln drive torque. Control adjustments are exercised on the fuel feed rate and exhaust damper position. They influence the various stages of heating, calcining, clinker formation, and cooling, with delays ranging from minutes to hours. In the burning process, surplus oxygen is controlled via fuel rate regulation. The airflow is constant. When the oxygen percentage is too high or too low, the fuel rate is adjusted.

High and *low* are linguistic values for the linguistic variable *oxygen percentage*. They can be represented as functions with values in the unit interval, that is, as fuzzy numbers. According to a human operator, a percentage of 1.38 is not high, is acceptable with the degree 0.98, and is low with the degree 0.47.

The variable "fuel feed rate" can take values Negative, Positive, and Zero. All the linguistic values can be modified by hedges like *very*.

From a human expert or from a textbook for cement kiln operators, one can obtain the following rules describing how the operator must adjust fuel rate under varying conditions characterized by oxygen in exhaust gases. The conditions and the control actions are specified in qualitative terms:

> IF oxygen is low, THEN fuel feed rate is medium negative.
>
> IF oxygen is acceptable, THEN fuel feed rate is zero.
>
> IF oxygen is high, THEN fuel feed rate is medium positive.

Sometimes the premise of the rules is a conjunction of different statements. For instance:

> IF kiln drive torque is acceptable
> AND
> free lime content is low
> THEN
> make a medium negative fuel rate adjustment
>
> IF kiln drive torque is acceptable
> AND
> free lime content is acceptable
> THEN
> make no fuel rate adjustment
>
> IF kiln drive torque is small
> AND
> free lime content is low
> THEN
> make a small positive fuel rate adjustment

IF kiln drive torque is negative
AND
free lime content is acceptable
THEN
make a medium positive fuel rate adjustment

A fuzzy controller for a cement kiln works by adjusting fuel flow (and other parameters) according to these rules. Each clause of each rule is tested in turn. The clause satisfied to the least degree by the actual conditions determines the degree to which the entire rule is satisfied. The actions recommended by each rule are weighted appropriately, and then the center of gravity of the resulting curb is used to determine the action that the controller will actually take.

AGGREGATE PRODUCTION PLANNING

Management sciences studies the performances of business corporations or entire industies by using simulation techniques to show how they respond to various conditions. A business operation is carried out by performing functions such as production and distribution. These functions are identified as the activities of a system and such elements as production level, inventory level, work force level, sales forecast, and demand are treated as system entities. To describe the behavior fully, we must consider a business as a whole so that all the interactions are seen. Typically, a production planning study will treat production capacity and change in work force as closely related.

The prime objective of a simulation study is to understand how the organization of a business affects its performance. The simulation aims to demonstrate the characteristic behavior of the system rather than to predict specific events.

Consider a production-inventory system whose state is described by a vector **x** with components

$$x_1 = \text{Production level}$$

$$x_2 = \text{Inventory level}$$

$$x_3 = \text{Work force level}$$

and whose input is described by a vector **u** with components

$$u_1 = \text{Change in work force}$$

$$u_2 = \text{Sales forecast}$$

$$u_3 = \text{Actual demand}$$

Usually the state equation describing the interrelations in time between these variables is nonlinear; that is, the relations are linked by nonlinear functions as follows:

$$x_1(k+1) = f[x_3(k), x_2(k), u_2(k)]$$
$$x_2(k+1) = x_2(k) + x_1(k) - u_3(k)$$
$$x_3(k+1) = x_3(k) + u_1(k)$$

where

$$u_1(k+1) = g[x_3(k), x_2(k), u_3(k)]$$

Nonlinear functions are difficult or impossible to identify and usually need another formulation, for example one based on IF-THEN rules (Rinks, 1982). For instance, we can get the following information (piece of knowledge) from a human expert:

IF $u_2(k)$ is HIGH and $x_2(k)$ is LOW AND $x_3(k)$ is HIGH
THEN
x_1 is HIGH

Primary terms such as Very high, High, Rather high, Sort of high, Average, Sort of low, Rather low, Low, Very low, can be defined as a continuum. Each term has a different meaning for each variable, and this meaning is fixed by giving lower and upper bounds. These bounds define the universe of discourse on which the linguistic values are defined.

A number of rules can be specified by transfering expertise from a human expert. For instance:

Rule	u_2	x_2	x_3	x_1	u_1
1	Sort of high	Low	High	High	Zero
2	Sort of high	Low	Average	Rather high	Positive Very big
3	Average	Average	High	Sort of high	Negative Rather big
⋮					

Many such IF-THEN rules can serve as the structure of a decision-making process.

Usually, the decision maker considers the requirements for a current period. The requirement could be

$$\text{Requirement}(k) = \text{Sales forecast}(k) - \text{Inventory level}(k)$$

The control rules can become

Requirement is high if production is high and inventory is average

Requirement is average if production is sort of high and inventory is sort of high
production is average and inventory is average
production is rather low and inventory is low

Requirement is low if production is low and inventory is average
production is sort of low and inventory is high

Any conditional statement is a syllogism of the form

$$\frac{\begin{array}{l}\text{IF } B \text{ is } b_1 \text{ THEN } A \text{ is } a_1 \\ B \text{ is } b_0\end{array}}{A \text{ is } a_0}$$

which is formed with a rule and a fact. From a rule in the knowledge base and a fact furnished by current measurments upon the system, we get a conclusion. The shape of the conclusion is given by an algorithm derived from the compositional inference, giving the membership function m of the conclusion b_0

$$m_{a_0}(z) = \min\left[m_{a_1}(z), \max \min\left[m_{b_0}(w), m_{b_1}(w)\right]\right]$$

Because we have a conjunction of clauses in the premise, our rules have the form

IF B is b_1 AND C is c_1 THEN A is a_1

ELSE IF B is b_2 AND C is c_2 THEN A is a_2

If B and C have the values b_0, c_0, then a_0, the inferred value of the conclusion variable, can be found again from the compositional inference rule.

AN AUTOMATIC TRAIN OPERATION

By train we shall understand a locomotive and a number of railway coaches (or wagons) joined together. The train is accelerated or decelerated to start or to stop. A human operator starts the train, keeps the speed below the specified limit, and decelerates to stop at the station. When the control command notch has a positive value, traction is controlled. The brake is controlled by negative values. At any moment the control command notch is a function of the velocity of the train, the maximum speed limit indicated by a signal, and many other parameters.

Human operators operate trains by evaluating such performance indexes as safety, comfort, stop accuracy, and so on. They have some rules of thumb or linguistic control strategies to follow. For instance:

For *acceleration*:
  For *safety*
    IF the speed exceeds the limit THEN select the max Brake notch
  For *saving energy*
    IF coasting can keep the scheduled running time
    THEN keep coasting
  For *shortening running time*
    IF the speed is far below the limited one
    THEN the power notch is selected
  For *riding comfort*
    IF the speed is in the predetermined allowance range
    THEN the control notch is not changed
  For *traceability*
    IF the train is about to go out of the allowance range
    THEN a $\pm n$ notch is selected so that the train
    stops at the target position

For *deceleration*:
  For *riding comfort*
    IF the train will stop in the predetermined allowance zone
    THEN the control notch is not changed
  For *shortening the running time*
    IF the train approaches the deceleration zone
    THEN the notch is changed to deceleration by degrees
  For *stop accuracy*
    IF the train will not stop within the predetermined zone
    THEN a $\pm n$ notch is selected so that the train will
    stop at the target position

These results can be applied as in the cement kiln control example. However, this method does not evaluate the results of the selected control command as human operators do. To overcome this disadvantage, we can apply a predictive approach (Myiamoto and Yasunobu, 1984). The control rule becomes

$$\text{IF [Notch is } N_i \text{ IMPLIES } x \text{ is } A_i \text{ and } y \text{ is } B_i]$$
$$\text{THEN Notch is } N_i$$

For *acceleration*:
IF [Notch is ($N(t)$ + Max Brake/2 IMPLIES safety is poor]
  THEN Notch is ($N(t)$ + Max Braking Notch/2)
IF [Notch is zero IMPLIES safety is good, comfort is good and energy is saved]
  THEN Notch is zero
IF [Notch is Power notch 7 IMPLIES safety is good, comfort is good, and traceability is low]
  THEN Notch is Power notch 7
IF [Difference of notches is zero IMPLIES safety is good and traceability is good]
  THEN Difference in notch is zero
IF [Difference in notch is n IMPLIES safety is good, comfort is good, and traceability is accurate]
  THEN Difference in notch is $n (n = \pm 1, \pm 2, \pm 3)$

For *deceleration*:
IF [Difference in notch is zero IMPLIES running time is not in deceleration zone and stop gap is good]
  THEN Difference in notch is zero
IF [Notch is zero IMPLIES running time is in deceleration zone and comfort is good]
  THEN Notch is zero
IF [Notch is 1 IMPLIES running time is in deceleration zone and comfort is good]
  THEN Notch is zero
IF [Difference in notch is n IMPLIES running time is in deceleration zone, comfort is good, and stop gap is accurate]
  THEN Difference in notch is n

The control command is selected by evaluating all the rules every 100 ms. Evaluation is based on predefining the terms "good," "poor," "low," "accurate," "saved," and so on. These terms are functions defined on different universes of discourse. For instance:

Poor safety is a function defined on two parameters:
  t_s time to reach the location where
    the maximum speed limit is lower
  T_s predicted time to reach this location
  Such a function can be $T_s/(T_s - t_s)$.

INVESTMENT DECISIONS

A predictive approach could be adopted also in one-stage actions. An investment operation is a one-stage, multichoice multicriteria decision problem. Human experts choose investments by evaluating performance indexes such as

$$i_1 = \text{Risk of losing capital}$$
$$i_2 = \text{Vulnerability to inflation}$$
$$i_3 = \text{Interest received}$$
$$i_4 = \text{Cash realizability}$$

An investment operation model could be formulated as

$$\text{The suitability of the investment } (k)$$
$$= \text{Function of } (i_1, i_2, i_3, i_4)$$

"Risk of losing capital" is an attribute with a clear, although not precise, meaning. It can intuitively have low or high values. These values can be defined on a continuum.

A computer-implemented system can realize an operation based on human expertise if this expertise i is captured in rules like the following:

Investment in commodity market implies risk is high plus
                                   inflation is sort of high
                                   interest is very high
                                   cash is fair

Investment in stock market implies risk is fair plus
                               inflation is fair
                               interest is fair
                               cash is sort of good

Investment in Gold or Diamonds implies risk is low plus
                                   inflation is sort of low
                                   interest is fair
                                   cash is good

Investment in real estate implies risk is low plus
                               inflation is very low
                               interest is sort of high
                               cash is bad

Investment in long-term bonds implies risk is very low plus
                                    inflation is high
                                    interest is sort of low
                                    cash is very good

The right side of each rule could be evaluated by adding the linguistic values of the variables. That rule that produces the highest fuzzy number on the real line is selected.

It is easy to see that the meaning of the words high and low is not the same in the context of different criteria like Risk and Interest. Clearly, High risk is a bad feature of an investment, and high interest is a good one. "Bad" and "good" are opposite evaluations, which, on a continuum scale, stay at opposite ends.

Because linguistic values like high can have different connotations, they are context dependent. *Context dependency* can be made explicit and exploited as a storage save advantage. The idea is to use the same term, which in the machine is a table or a vector, for different words having the same meaning. If High risk and Low interest have the same meaning, that of being undesirable, then they can be specified by the same linguistic value.

Let us consider again the fuzzy numbers used previously to define different values for RISK. They can be used now to define the meaning of the following vocabulary:

Fuzzy Number	Meaning for Risk	Meaning for Inflation	Meaning for Interest	Meaning for Cash
(0,0,0,2)	Very high	Very high	Very low	Very bad
(0,1,0,2)	High	High	Low	Bad
(2,2,2,2)	Sort of high	Sort of high	Sort of low	Sort of bad
(5,5,2,2)	Fair	Fair	Fair	Fair
(8,8,2,2)	Sort of low	Sort of low	Sort of high	Sort of good
(9,1,2,0)	Low	Low	High	Good
(10,10,2,0)	Very low	Very low	Very high	Very good

Given an internalized vocabulary, the machine can evaluate the five rules representing its knowledge base. This can be done by considering the right sides of the rules, which become

    High   + Sort of high + Very high + Fair

    Fair   + Fair   + Fair   + Sort of good

    Low   + Sort of low + Fair   + Good

Low   + Very low   + Sort of high + Bad
Very low + High   + Sort of low + Very good

The operation of addition is executed according to the rules of fuzzy arithmetic.

Because all our linguistic values were defined on the continuum [0, 1, 2, 3, 4, 5, 6, 7, 8, 9, 10], these values are tables specifying the membership of these numbers to each value. Let

$$\text{High} = [\text{High}(i)/i, 0 < i < 10]$$
$$\text{Low} = [\text{Low}(i)/i, 0 < i < 10]$$

then

$$\text{High} + \text{Low} = [\min(\text{High}(i), \text{Low}(j)/[i+j]), 1 < i < 10]$$

What does this definition mean computationally? To compute the degree of membership of, say, 8 in High + Low, we have to examine all of the possible ways that two integers taken from the set $\{0, 1, 2, 3, 4, 5, 6, 7, 8, 9, 10\}$ can sum to 8, and examine the degrees of membership of these pairs. Thus, if the degree of membership of 8 in High + Low was x, then x would be computed as follows:

$$x = \max[\min[\text{High}(1), \text{Low}(7)], \min[\text{High}(2), \text{Low}(6)],$$
$$\min[\text{High}(3), \text{Low}(5)], \min[\text{High}(4), \text{Low}(4)]$$
$$\min[\text{High}(5), \text{Low}(3)], \min[\text{High}(6), \text{Low}(2)]$$
$$\min[\text{High}(7), \text{Low}(1)], \min[\text{High}(0), \text{Low}(8)]$$
$$\min[\text{High}(8), \text{Low}(0)] \qquad\qquad\qquad]$$

Each of the minimum operations computes one of the degrees of membership of 8 in the fuzzy set High + Low. We then take the greatest such degree of membership to be the degree of membership of 8. Notice that High + Low is a fuzzy number over the set of integers from 0 to 20.

The computational burden of the implementation of this algorithm could be considerable. The use of sampled membership distributions generates a considerable increase in the finite discrete support of the result. Furthermore, the membership distribution of the result is no longer sampled at regular intervals. The solution of this problem consists of using a parameter-based representation for fuzzy sets (Tong and Bonissone, 1980).

INFERENCE RULES AND FUZZY STATISTICS

Knowledge engineering, as the name implies, is concerned with the problem of managing an accumulated body of truths and facts integrated into computer systems to solve complex problems normally requiring a high level of human expertise.

According to this definition, knowledge engineering is simply subfield of artificial intelligence, which is a subfield of computer science, and appeared on the scene when researchers were convinced that world knowledge is behind any intelligent behavior.

In more technical jargon, artificial intelligence is possible when a machine internalizes a model of the external milieu. This is true for the case of expert systems, no matter whether they are based on symbolic manipulations of production rules or on semantic manipulations of linguistic variables. An expert system can pose and answer questions relating to information borrowed from human experts and stored in a knowledge base.

A linguistic controller can act automatically according to information borrowed from human experts and stored in a knowledge base along with a dictionary giving the meaning of words.

An intelligent knowledge-based system uses inferences. The knowledge-based approach to system design replaces the software paradigm based on the idea of a program with a new one centered around an inference engine.

The distinctive characteristic of any knowledge-based system is that its processes are state driven rather than hard coded. Decisions about how to process data are part of the knowledge of the system. In other words, by internalizing knowledge as a model of the world, a machine becomes intelligent.

Inference is the act of concluding from premises or evidence. Consider the following piece of knowledge:

IF price is low THEN profit is negative.

This is a rule, or an implication, a relation that can be written as $A \rightarrow B$, where A is called the *antecedent* or *premise* or *condition*, and B is called the *consequent* or *conclusion* or *action*.

The inference engine is a mechanism that combines the rules. It can be antecedent driven or consequent driven. In an antecedent-driven system, the occurrence of an antecedent triggers a rule to infer its consequence. For instance, the system with two rules

Low price → High profit
High profit → Good shares

triggered by the fact "low price" introduced from outside by a user, can infer "good shares."

In a consequent-driven system, the existence of a consequent is verified by confirming that the antecedent is present. For instance, if the user is coming with the query Are the shares good? the system takes the query as a goal, inspects the rules, finds the second rule, and sees that its condition is "high profit." Now, "high profit" becomes the goal, the system inspects the rules, finds the first one, and sees that its condition is "low price." Because there are no other rules, the system asks the user to verify that the price is low. If so, the answer is Yes, with the meaning that the shares are good.

A rather simple control structure assumes that the system has the overall goal of confirming or discarding a hypothesis on the basis of the facts (evidence) introduced by the users.

Consider an antecedent-driven system loaded only with the rule $A \rightarrow B$. If the user introduces a fact A in the system, the inference is based on the syllogism

$$\begin{array}{ll} \text{Major premise} & A \rightarrow B \\ \text{Minor premise} & \underline{A} \\ \text{Conclusion} & B \end{array}$$

This is a way of saying that the condition of the rule matches the state of the inference engine, and the action part of the rule is exercised. The process can be repeated when this action part has the role of the fact and, in turn, matches the state of the inference engine, which is a condition of a new rule, in a chain reaction.

Clearly, the production rules must be executed in an all-or-nothing manner. Rules are executed if their left side matches; otherwise rules are not executed. For a production system to make use of a variable measured on a numeric scale, the scale must be divided into regions, each region having its rule. In other words, there is no concept of "closeness."

Such a concept is possible in a framework based on the acceptance of uncertainty. Various schemes have been tried, some quite successfully, that avoid the all-or-nothing approach.

Consider the rule

IF it rains THEN take an umbrella.

Clearly, the rule is immediately executed if the fact from outside is "it rains." But suppose the fact is

Weather report indicates rain

In the first case the inference is given by the syllogism

> Rain → Umbrella
> Rain
> ―――――――――
> Umbrella

which works. In the second case the inference is given by the syllogism

> Rain → Umbrella
> Predicted Rain
> ―――――――――
> Umbrella?

which does not work. How can we use the syllogistic scheme in this case? An answer is offered by mathematical statistics.

Mathematical statistics is the study of how to deal with facts by means of probability models. It grew out of methods for treating data that were obtained by some repetitive operations, such as those encountered in games of chance.

The outcome of a rule is uncertain, but if repeated a number of times one may be able to construct a probability model for it and make decisions by means of it. Experience indicates that many repetitive operations behave as though they occurred under essentially stable circumstances.

The mathematical model that statisticians select for a repetitive operation is usually one that enables them to make predictions about the frequency with which certain results can be expected to occur. For instance, in our example if we know the probability of raining, we can work out the probability of the rule in the light of evidence.

The whole essence of this approach, known as Bayesian, rests on the belief that for everything there is a probability to be true. Now, given a prior probability about some rule, there must be some evidence we can call on to adjust our beliefs on the matter. Given relevant evidence, we can modify this prior probability to produce a posterior probability of the same rule, given the new evidence.

Consider the rule Rain → Umbrella. We know roughly what the prior probability of this rule is, say $P(R)$, and we want to know the probability for the new evidence "weather report indicates rain." $P(R/E)$ is the posterior probability of the same rule R given evidence E. By definition

$$P\left(\frac{R}{E}\right) = \frac{P(R \text{ and } E)}{P(E)}$$

$$P\left(\frac{E}{R}\right) = \frac{P(E \text{ and } R)}{P(R)}$$

Therefore

$$P\left(\frac{R}{E}\right) = \frac{P(E/R)P(R)}{P(E)}$$

We start off with $P(R)$, a value held in the knowledge base, and we calculate $P(R/E)$, having asked about facts. If the fact is "weather report indicates rain," we can forget the original $P(R)$, valid for "it rains" and use instead $P(R/E)$. This process can be repeated time and again as new evidence comes from the users, for instance for facts as

The spouse says it will rain.
The sky is overcast.
You just bought a new umbrella.
The sky is clear.

In our framework each fact is represented by a probability, and matching the facts with the rule takes place according to the syllogism

Probability(rain) → Probability(umbrella)
Probability(predicted rain)
―――――――――――――――――――――
New probability(umbrella)

The difficulty was reduced by having the same nature for the major and minor premises, namely a number.

The number is the fundamental tool in the inference mechanism because the statistician looks at the probability as an idealization of the proportion of times that a certain result will occur in repeated trials. Although interpreted as such, the probability is seldom treated as a measure of an individual's betting odds. In applying probability techniques to such problems, however, we must realize that their reliability is heavily dependent on the realism of the individual betting odds. To be on the sure side, we usually do not say "the probability is .5," but "the probability is high." In other words, a probability can take linguistic values.

The syllogism now becomes

Linguistic value for p(rain) → linguisitc value for p(umbrella)
Linguistic value for p(predicted rain)
―――――――――――――――――――――
Linguistic value for p(umbrella)

Clearly, the major and minor premises have the same nature, and we are still in the framework of the classical matching of a rule with a fact. But again they have different names, and if we want to avoid an all-or-nothing manner of execution something has to be done in order to accept "closeness." A new transformation is necessary, and this is done by modeling them as fuzzy sets.

Consider the syllogism

High probability(rain) → High probability(umbrella)
Low probability(rain)
―――――――――――――――――――――――――――――――
Medium probability(umbrella)?

We can model High as a function defined on the unit interval because a probability can take values only in this interval. This function has to allow a continuum of choices. Clearly, the probability .9 is high, so the value of this function is maximal. If we again take the unit interval as the codomain of the function, we can say that the probability .9 belongs to High with membership degree 1.0. It is not easy to say that a probability of .6 is high. One can say that this probability is partially high. We can evaluate the degree of highness as .7, and so on. In this way the vagueness of the term High can be captured mathematically and dealt with in an algorithmic fashion. We say that the linguistic value High is a fuzzy set defined by the characteristic function

$$\text{High}: [0.1] \to [0.1]$$

Replacing probabilistic models with possibilistic ones is a fascinating enterprise. No less fascinating is the attempt to combine them.

This book introduces the concept of a fuzzy random variable as a new way of describing imprecise information. More specifically, such variables will be used to describe fuzzy data associated with the outcomes of a random experiment. This concept generalizes that of a random set as well as that of a random variable and seems to be well suited as a model of those situations where a mixture of probabilistic and possibilistic information arises.

A fuzzy random variable is a function

$$X \to F(R^p)$$

where X is a sample space and $F(R^p)$ denotes all fuzzy sets of the Euclidean space R^p—that is, all functions $R^p \to [0,1]$.

We will show how to develop a comprehensive theory of fuzzy random variables, including a strong law of large numbers and a central limit theorem for independent fuzzy random variables. We also introduce the concept of normality for fuzzy random variables, which is useful for making estimations based on imprecise data.

Suppose that we observe a random sample of fuzzy variables whose common distribution depends on some unknown fuzzy parameters. The available information is partly statistical and partly fuzzy. We will investigate the general problem of making inferences about these fuzzy parameters.

Another important problem that arises is that of finding confidence regions for fuzzy parameters. One can investigate various properties of these regions, including admissibility, unbiasedness, and asymptotic convergence.

To obtain meaningful results, we will restrict the range of fuzzy random variables. More specifically, the imprecise data should be described by fuzzy sets satisfying certain properties, such as the compactness of their levels. We need the *theorem of representation* as a key tool for all our development.

READINGS

The first application of fuzzy set theory to derive a linguistic control strategy for a steam engine and boiler combination using the heuristic rules stated by a human operator was reported by

E. H. Mamdani, and S. Assilian, 1975, An experiment in linguistic synthesis with a fuzzy logic controller, *International Journal of Man-Machine Studies* 7:1-13.

Similar successes in applications on processes that obey complex physical laws were reported by

E. H. Mamdani, and J. King, 1977, The application of fuzzy control systems to industrial processes, in *Fuzzy Automata and Decision Process*, M. Gupta, G. Saridis, and B. Gaines (eds), Amsterdam: North-Holland.

W. J. Kickert, and H. Van Nauta Lemke, 1976, Application of a fuzzy controller in a warm water plant, *Automatica* 12:301-308. J. J. Ostergaard, 1977, Fuzzy logic control of a heat exchanger process, in *Fuzzy Automata and Decision Processes*, M. Gupta, G. Saridis, and B. Gaines (eds.), Amsterdam: North-Holland.

A detailed account of rule-based techniques for the control of industrial processes can be found in

R. M. Tong, 1977, A control engineering review of fuzzy systems, *Automatica* 13:559-569.

E. H. Mamdani, J. J. Ostergaard, and E. Lembessis, 1984, Use of fuzzy logic for implementing rule-based control of industrial processes, in *Fuzzy Sets and Decision Analysis*, H. J. Zimmerman, L. A. Zadeh, and B. Gaines (eds.), Amsterdam: North-Holland.

Of particular interest are the following works:

L. P. Holmblad, and J. J. Ostergaard, 1982, Control of a cement kiln by fuzzy logic, in *Fuzzy Information and Decision Processes*, M. Gupta and E. Sanchez (eds.), Amsterdam: North-Holland.

Cement kilns exhibit time-varying nonlinear behavior and relatively few measurements are available. Consequently, automatic control is usually restricted to a few simple control loops on secondary variables, leaving the responsibility for the control of primary variables to the kiln operators. The usual approach when seeking a higher degree of automatic control of the kiln operation has been to establish a mathematical model of the kiln process. However, experience indicates that mathematical process models either become too simple to be of any practical value or too comprehensive to possess any general applicability.

There is a striking contrariety between the difficulties encountered in establishing adequate mathematical descriptions of the kiln process, on the one side, and the relative ease by which human operators can be trained to become skilled. In this connection it was observed that a mathematical model approach builds on absolute numerical quantities, whereas human operators seem to think and act according to approximate relationships involving vaguely defined, linguistic quantities like "high," "small," "OK."

This paper describes a program marketed by F. L. Smidth & Company of Copenhagen, Denmark, which specializes in operating cement kilns. The program controls the temperature of the fires, the rate of fuel flow, and the operation of the exhaust fans and dampers. By using fuzzy logic, it avoids the abrupt changes that might result from the either-or, all-or-nothing judgements inherent in classical logic. That way the kiln's machinery functions smoothly with fewer sudden stops or speed-ups that might damage the equipment and affect the quality of the cement.

The fuzzy control system is incorporated in F. L. Smidth's computerized process monitoring system, FLS Supervision, Dialogue and Reporting (SDR) System. The SDR system is designed for use in the cement industry for monitoring, operator communication, and reporting tasks as well as for general control in conjunction with other equipment.

The SDR system is designed around a minicomputer to which is connected a graphical operator color screen with keyboard, graphic system console, and printer. Typical computer size is 256K, and external storage (disks) has been avoided.

The SDR system contains a programming language and a program interpreter, both developed for the specification and execution of control strategies. Program language and interpretation, called fuzzy control language (FCL) are similar to other interpretation-based languages such as BASIC and APL. The FCL system requires approximately 32K of the SDR computer storage capacity.

A control strategy consists of one or more FCL programs. Consider the verbal control strategy:

> When CO_2 is registered and exhaust gas temperature is low or around the required level, increase air flow. If some smoke chamber temperature is high, it is better to slightly reduce the coal feed. To allow for process reaction, wait some time before taking any new action.

An analysis of this strategy shows that it consists of two control rules specifying the control action, depending on the actual level of the smoke chamber temperature. The strategy also demands a certain interval between successive actions.

If the variables CO and TEMP are introduced, the two rules can be formulated in FCL as

```
IF HIGH(CO) AND (OK(TEMP) OR LOW(TEMP))
   THEN
     LPOS(DS1), ZERO(DF1)

IF HIGH(CO) AND HIGH(TEMP)
   THEN
     SNEG(DF1), ZERO(DS1)
```

where DS1 and DF1 are variables for adjusting louver damper and coal feed, respectively. LPOS and SNEG are abreviations for large positive and small negative.

Each FCL program has a name, and a time interval between each execution is specified. For input variables CO and TEMP and for output variables DS1, DF1, it is necessary to define connections to the data base of measurements and set points of the SDR system which contain measurement values and set points for coal feed and louver damper position. A scaling interval for CO and TEMP must be specified to determine which values are regarded as low, ok, or high.

S. Myiamoto, and S. Yasunobu, 1984, Predictive fuzzy control and its application to automatic train operation systems, paper presented at the *Fuzzy Information Processing '84*, Hawaii.

In recent years many automatic control systems have been developed that use a microcomputer instead of human operators. However, most of them achieve a control quality below the level of a skilled human operator. This difference is due to nonlinear time-variant behavior inherent in real processes with multiperformance indexes. The conventional automatic control model linearizes process behavior and reduces the performance indexes to one. The fuzzy control algorithm has been used to implement directly heuristic control policies expressed linguistically in order to automate those complex and poorly defined processes. However, conventional fuzzy control does not evaluate the results of selected control command the way human operators do. To overcome this problem, the authors proposed predictive fuzzy control.

This paper describes a controller developed by the Systems Development Laboratory, Hitachi Ltd., Japan. The developed fuzzy controller was compared with a conventional controller. The results of the simulation were as follows:

There is no difference in controllability, but the change of frequency of notches with a fuzzy regulator is about half of the conventional controller. That is, the riding comfort was improved.

The stop gap of the fuzzy controller was smaller. That is, the robustness and accuracy of the fuzzy controller was greater.

The fuzzy controller can operate trains with an energy saving more than 10% above that of the conventional controller and shorten the running time.

More details can be found in

M. Sugeno (ed.), 1985, *Industrial Applications of Fuzzy Control*, Amsterdam: North-Holland.

D. B. Rinks, 1982, The performance of fuzzy algorithm models for aggregate planning under differing cost structures, in *Fuzzy Information and Decision Processes*, Amsterdam: North-Holland.

One of the most important decisions that must be considered in manufacturing is capacity. One facet of this decision is concerned with scheduling equipment and a work force in addition to managing inventories. In operations management the process of simultaneously analyzing work force size, production rate, and projected size of inventories is known as aggregate planning and scheduling.

Although a multitude of mathematical models of aggregate production planning exist in the literature, implementation of any of these models is practically nonexistent. One explanation for the lack of use of the mathematical models is that managers believe that the techniques and their associated cost models simply do not adequately represent the realities of their operations. Apparently, most managers have decided to use their own heuristics, or intuitive decision rules, to be erratic in the operation of their judgmental models. Thus, lack of consistency in applying judgmental models, rather than the form of the heuristic, is often suggested as the reason for the poor decisions generated by judgmental processes. It follows then that a consistently applied formal model of the manager's judgmental process should outperform the manager's actual judgmental process.

In their attempt to model human decision making, analysts construct heuristic programs that seek to replicate the thought processes of humans. The first step in such procedures is usually to solicit the protocol of the decision maker, a process whereby the decision maker verbalizes thought processes while actually making decisions.

Rinks investigates the plausibility of using linguistic variables to directly translate a decision maker's protocol into a computer-implemented model. To test the performance of the aggregate planning fuzzy algorithm, Rinks used the classic Holt-Modigliani-Muth-Simon paint factory data. His results proved that the fuzzy algorithm can closely approximate the results of much more powerful optimizing techniques.

R. M. Tong, and P. P. Bonissone, 1980, A linguistic approach to decision making with fuzzy sets, *IEEE Transactions on Systems Man and Cybernetics* **SMC-10:**716–732.

R. M. Tong, and P. P. Bonissone, 1984, Linguistic solutions to fuzzy decisions problems, in *Fuzzy Sets and Decision Analysis*, H. J. Zimmermann, L. A. Zadeh, and B. Gaines (eds.) Amsterdam: North-Holland.

Previous attempts to solve fuzzy decision problems have produced numerical rankings of the alternatives. The authors believe that this is unnecessary, because in situations where fuzzy sets are a suitable way of representing imprecision the final choice could by fuzzy. They assume that in order to choose from among a set of alternatives, it is enough to have fuzzy information about the "suitability" of each of them. Suitability is interpreted as a measure of the ability of an alternative to meet decision criteria. The concept is the fuzzification of the idea of a rating. The first step in their procedure is to compress suitability information into a single fuzzy set that represents the decision. Selection is made via the concept of dominance. Because the universe of discourse on which the elements of the suitability fuzzy sets are drawn is the universe of numbers, the linearly ordered real line, the "best" alternative is the one defined by a suitability fuzzy set with a peak that lies to the right of all others. Therefore, the decision is finally made in a numerical form, proving that *Defendit numerus* (the numbers protect) is the maximum of the wise.

E. Charniak, and D. McDermott, 1985, *Introduction to Artificial Intelligence*, Reading, Mass.: Addison-Wesley.

This beautiful, academic text offers a painless way to learn about machine intelligence. The authors define it as the study of mental faculties through the use of computational models. The fundamental working assumption or central dogma of artificial intelligence is that what the brain does may be thought of at some level as a kind of computation. Exactly what kind of computation is the subject of this book. It is centered around the key idea of internal representation of facts and rules. A representation is considered to be a stylized version of the world.

Processes that use internal representation require the ability to draw inferences. To draw an inference is to come to believe a new fact on the basis of other facts. There are many kinds of inferences. The best understood is deduction, which is logically correct inference. This means that deduction from true premises is guaranteed to result in a true conclusion. The standard way to characterize deduction is by using a system called the predicate calculus, a language for expressing propositions and rules for how to infer new facts (propositions) from those we already have.

Inference is predicate calculus's strong point. The initial facts that drive the deductions are called axioms. These are the things to assume and are to be distinguished from the things we deduce, which are theorems. In mathematics we normally think of the axioms as being set out in advance and never changed. In artificial intelligence there is a different view. Predicate calculus is used to represent information about the outside world. This information will become further axioms for our system. In general, the new facts will not be deducible from previous facts, and since all facts are either axioms or theorems, the new facts must be axioms. Thus, in contrast to mathematics, our set of axioms will change over time.

To allow us to deduce new facts from the axioms, the predicate calculus has rules of inference. The most famous is *modus ponens*, which says that given a rule (if p then q) and a fact (p), we are allowed to deduce another fact (q). A second kind of inference is called *abduction*. Abduction is the process that generates explanation. In other words, given a rule (if p then q) and a fact (q), we can infer another fact (p). Abduction allows false conclusions; medical diagnosis is an example.

Like deduction, abduction requires that we find pertinent facts and apply them to infer a new fact. However, abduction, having more than one possible answer, requires something else. The extra required step is to decide which answer to select. The best one can do is to find the most probable hypothesis. To make decisions of this sort in face of uncertainty, one must weigh evidence. To do this, we must know how to combine pieces of evidence into a final conclusion. Statistics offers ways of doing such things. In other words, abduction is probable inference.

The book has a chapter devoted exclusively to abduction uncertainty and expert systems. An expert system is defined as a rule-based program for doing a task that requires expertise. Most expert systems have performed abductive tasks, such as medical diagnosis. Others are better classified as problem-solving programs acting on a working memory (a set of data structures representing the current state of the system), a production memory (a set of rules), and a rule interpreter that applies the production rules to the working memory.

An important issue in designing a production system is conflict resolution, because a production system achieves sequential behavior without being sequenced.

P. H. Winston, 1984, *Artificial Intelligence*, 2nd ed., Reading, Mass.: Addison-Wesley.

Any procedure for computing certainty factors must embody answers to three questions. First, how are the certainties associated with the rule's antecedents to be combined into the rule's overall input certainty? Second, how does the rule itself translate input certainty into output certainty? Third, how is a fact's certainty determined when the consequents of several antecedent-consequent rules argue for it, requiring the computation of a multiple argued certainty? The simplest idea is that of using the product of antecedent certainties to get the overall certainty of the input. The idea is derived directly from the notion that the probability of a joint event is the product of the probabilities of the participating events, as long as the participating events have no influence on one another. In a coin toss, for example, the probability of turning up two heads in a row is the square of the probability of turning up one head on one toss. The antecedents of a rule often are dependent events. What we are computing is analogous with a probability, but the combining formula is not justified by the probability theory. None of these procedures are completely satisfactory. We are looking at what we can do now, not at what we want to do.

This book does not mention fuzzy logic but the active practitioners in the filed were aware of it from a long time, as for instance in

R. Forsyth (ed.), 1984, *Expert Systems. Principle and Case Studies*, London: Chapman and Hall.

This is a contributed volume explaining the concepts behind expert systems, viewed as computer systems that encapsulate specialist knowledge about a particular domain of expertise and capable of making intelligent decisions within that domain, as was the case in medical diagnosis, geologic exploration, organic chemistry, and fault-finding in different machines.

It is precisely the expert systems that have given rise to a set of knowledge engineering methods constituting a new approach to the design of high-performance software systems. An important chapter of knowledge engineering considers the problem of inference—how to use approximate reasoning strategies to arrive at a good estimate of the truth with uncertain data and imperfect rules. This is an area where expert systems designers have pioneered a breakthrough in computing practice: knowledge-based systems do not rely on zero-defect software engineering.

Fuzzy logic, certainty factors, and Bayesian statistics are carefully examined with a debate over the frequency interpretation of unrepeatable events and the relationship between degrees of relational belief and probability.

The reader who wants to go deeper into this topic has to see the opinion of the founder of fuzzy logic as for instance in

L. A. Zadeh, 1983, The role of fuzzy logic in the management of uncertainty in expert systems, *Fuzzy Sets and Systems* 11:199–227.

The author starts with the observation that management of uncertainty is an intrinsically important issue because much of the information in a knowledge base is imprecise, incomplete, or not totally reliable. In fact, most of the facts and rules in expert systems contain fuzzy predicates, as for example in PROSPECTOR, an existing expert system where uncertainty is dealt with through a combination of predicate logic and probability-based methods. In the rule

> IF abundant quartz sulfide veinlets with no
> apparent alteration halos
> THEN alteration favorable for the potassic stage

almost everything is a fuzzy predicate.

Fuzzy logic provides another framework for the management of uncertainty, because in contrast to traditional logical systems, its main purpose is inference from imprecise rather than precise knowledge.

A generalized version of modus ponens—the compositional rule of inference—stated in the form of the syllogism

$$H: \text{ IF } x \text{ is } F \text{ THEN } Y \text{ is } G$$
$$\frac{x \text{ is } F^*}{y \text{ is } H \circ F^*}$$

where "$\circ$" stands for composition, differs from its classical version in two respects. First, F^* is not required to be identical to F. Second, the predicates

F, G, F^* are not required to be crisp. It can be readily verified that when $F = F^*$ and the predicates are crisp, $H \circ F^*$ reduces to G.

For qualified propositions the problem of inference becomes much more complex. In a canonical from a qualified proposition may be expressed as

$$QA\text{'s are }B\text{'s}$$

For example,

Most cars are unsafe.

Zadeh introduces the intersection/product syllogism expressed in the scheme

$$\frac{Q_1 A\text{'s are }B\text{'s}}{Q_2(A \text{ and } B)\text{'s are }C\text{'s}}$$
$$(Q_1 \circ Q_2)A\text{'s are }(B \text{ and } C)\text{'s}$$

in which A, B, and C are fuzzy predicates, and $Q_1 \circ Q_2$ is a fuzzy number. For example,

$$\frac{\text{Most students are single}}{\text{Most single students are male}}$$
$$(\text{Most})^2 \text{ of students are single and male}$$

The reader interested in the problem of quantifiers included in production systems can find a detailed treatment in

A. L. Ralescu, 1986, A note on rule representation in expert systems, *Information Science* **38**:193–203.

The author considers typical rules in an expert system of the form

(a) IF X is A, THEN Y is B.
(b) IF X_i is A_i ($i = 1, 2, \ldots, n$), THEN Y is B.
(c) IF Q of X_i are A_i ($i = 1, 2, \ldots, n$), THEN Y is B.

X, Y, X_i are variables with values in U, and A, B, A_i are fuzzy subsets of U, and Q is an imprecise quantifier. Statement (b) can be viewed as a special case of (c), with Q having the value "all", and (c) can be written

IF Q of X_i is A_i ($i = 1, 2, \ldots, n$) hold, THEN Y is B.

If Q is represented as a fuzzy set, a formula is derived to calculate the conclusion of the syllogism.

This paper is a good example of the recent effort to solve a basic problem in the design of expert systems: how to equip them with a computational capability for evidence transmission.

Other details can be found in

C. V. Negoita, 1985, *Expert Systems and Fuzzy Systems*, Menlo Park, Calif: Benjamin/Cummings.

This book deals with semantic manipulations versus symbolic ones. When inference procedures are augmented with mechanism for plausible reasoning, conclusions are drawn from facts that seem to be correct. When inference procedures are augmented with mechanisms for approximate reasoning, conclusions are drawn from facts considered to be fuzzy. The treatment of fuzziness is a critical issue in knowledge representation. To say that a word is fuzzy is to say that there is no definite answer as to whether the word applies to something. The indeterminacy is due to an aspect of the meaning of the word rather to the state of our knowledge. In expert systems based on semantic manipulation and approximate reasoning, the emphasis is on fuzziness viewed as a intrinsic property of natural language.

An evident advantage of the fuzzy set approach is the possibility of representing numeric and linguistic variables in a uniform way and of using a sound formalism to handle them.

If we represent facts as objects and rules as morphisms the mathematical theory of categories is a good language for describing the mechanism of evidence combination. The difference between plausible and approximate reasoning becomes the difference between the categories on which we model the facts and the rules.

The practical advantage of using a catagorical approach is that each category has its own predicate calculus, and these predicate calculi are related to the nature of objects, therefore with the kind of representation adopted in the system. Relating representation and inference, we can get a deeper understanding of the fragile processes of knowledge engineering.

The reader interested in the categorical approach can find more details in

L. N. Stout, 1984, Topoi and categories for fuzzy sets, *Fuzzy Sets and Systems* **12**:169–184.

Topoi are of interest to logicians because they have an internal logic that allows higher-order theories to be modeled in them. Since almost all of mathematics can be phrased in higher-order theories, this means that mathematics can be made in a topos. Fuzzy mathematics use fuzzy sets for the same kind of modeling, though the logic of fuzzy sets is more external and the higher-order logic is not inherent.

Let H be a completely distributive lattice, and hence a Heyting algebra. The category of fuzzy sets, Set(H), and the topos of sheaves, Sh(H), are interconnected by pairs of adjoint functors between them. Each of the categories has a predicate calculus. These predicate calculi are related by the functors between them. Changing the base lattice gives rise to several functors that preserve or reflect specific kinds of statements in the predicate calculi.

This paper gives details of the categories, the predicate calculi attached to them, the functors between the categories, and the preservation properties of those functors.

For those readers interested less in mathematical logic and more in the philosophy of logic, we recommend

P. Achinstein 1983, (ed.), *The Concept of Evidence*, New York: Oxford University Press.

The selections in this volume represent a broad spectrum of theories of evidence and include material critical of rival viewpoints. They provide the reader with an opportunity to study seminal discussions in the field and the ensuing controversy. Perhaps the most influential article is by Rudolf Carnap. Interested in the logical foundation of probability, Carnap says that a classificatory concept of evidence can be defined only in terms of a quantitative one: the degree to which evidence e confirms hypothesis h. Carnap writes quantitative confirmation statements as $c(h, e) = r$, which means that the degree in which e confirms h is the number r. Carnap's c-function obeys the standard axioms of the probability calculus, so r is some number between 0 and 1.

Achinstein notes that a definition of this sort is offered by many writers, but despite its widespread acceptance it cannot possibly be correct if "evidence" and "probability" are being used as they are in ordinary language. For one thing, it does not require that evidence be true, and this seems to be necessary for evidence. In other words, increasing the evidence of a hypothesis is not the same as increasing its probability.

Achinstein states that we have to consider the general relationship between probability statements and background information. What view we take of this relationship will determine what evidence statements we can assert. According to Achinstein, the background information must be incorporated into the probability statement itself. This is achieved when we use linguistic values for probabilities, as in the statement

> The probability that John is sick, given that he has fever, is *high*.

Such a statement might be defended by appeal to the empirical fact that

> In *most* cases people with (that kind of) fever are sick.

The relation between probability and fuzziness has recently been made the object of intensive research. More details can be found in

A. L. Ralescu, and D. A. Ralescu, 1984, Probability and fuzziness, *Information Sciences* **34**:85-92.

The authors start from the fact that it is now generally accepted that probability theory and the theory of fuzzy sets are both used to study inexactness, the former as a model of statistical inexactness (due to the occurrence of random events) and the latter as a model of inexactness due to human judgment. These two approaches are not contradictory; neither includes the other, nor is one more general than the other. There is nothing to reconcile in these two approaches.

The main purpose of this paper is to show that methods of probability theory and of fuzzy set theory can be combined to model different sources of inexactness. Finally, it is shown that this combination makes possible the assessment of credibility in expert systems, which infer from vague premises according to typical rules of the form

IF X is A, THEN Y is B (with confidence factor c).

X and Y are variables, A and B are ordinary sets, and the confidence factor c is a conditional probability:

$$c = \text{probability}(Y \text{ in } B | X \text{ in } A)$$

If A is a fuzzy set instead of an ordinary set, the preceding probability will not make sense, and the confidence factor should be interpreted as a fuzzy number. Then X becomes a fuzzy random variable rather than an ordinary variable.

An issue of fundamental importance in the theory of expert systems is "credibility." How valid are the conclusions drawn from imprecise premises? It is important to observe that the statistical analysis based upon fuzzy random variables can help us to cope in a more systematic way with this issue.

Chapter 3

CONCEPTS OF FUZZY SET THEORY

FUZZY SETS AND FUZZY RELATIONS

The concept of fuzzy set, also associated with the term "graded membership," has been considered as a model for inexact, vague statements about the elements of an ordinary set. More specifically, in phrases like "large numbers," "tall men," "young children," "approximately equal to 10," the attribute, although common in the natural language, is not tractable by the methods of classical set theory or probability theory.

First, there is undecidability about membership or nonmembership in a collection of objects. Second, there is nothing random in the concepts in question.

An ordinary set or, more precisely, a subset of a given, total set X, can always be identified with a binary valued function $f_A: X \to \{0, 1\}$, called its characteristic function, and defined as

$$f_A(x) = \begin{cases} 1 & \text{if } x \in A \\ 0 & \text{if } x \notin A \end{cases}$$

The fact that subsets of X and characteristic functions are identifiable can be easily explained. Let $P(X)$ denote all subsets of X:

$$P(X) = \{A \mid A \subseteq X\}$$

$P(X)$ is called the power set of X. Let 2^X denote all functions from X into $\{0, 1\}$:

$$2^X = \{f \mid f: X \to \{0, 1\}\}$$

There is a one-to-one and onto correspondence between $P(X)$ and 2^X via the characteristic function $A \to f_A$. This correspondence preserves the usual operations with sets such as union, intersection, and complement. More exactly,

$$f_{A \cup B} = \max(f_A, f_B)$$
$$f_{A \cap B} = \min(f_A, f_B)$$
$$f_{\bar{A}} = 1 - f_A$$

The basic idea of a fuzzy set is very simple. Since there is undecidability concerning membership and nonmembership, one can consider a function defined on X whose values are in the unit interval $[0, 1]$ instead of just 0 and 1. Such a function measures the degree of membership and is called a membership function. More formally, we have

Definition

A fuzzy subset of X is a function

$$u: X \to [0, 1]$$

If the total set is clearly specified, we sometimes use the term "fuzzy set" instead of "fuzzy subset." To stress the point again, the function u describes an imprecise statement about the elements of X. For every $x \in X$ the value $u(x)$, $0 \leq u(x) \leq 1$, represents the membership degree of x in X—that is, the degree to which x satisfies that imprecise statement.

The collection of all fuzzy subsets of X will be denoted by $F(X)$ or by $[0, 1]^X$. Thus

$$F(X) = \{u \mid u: X \to [0, 1]\}$$

Clearly $F(X)$ extends $P(X)$, since every ordinary subset of X, called, in this context, a crisp set, is also a fuzzy subset of X. Therefore $P(X) \subseteq F(X)$.

Because of this inclusion, it is important that concepts and properties related to fuzzy sets constitute a generalization of the corresponding concepts and properties related to ordinary sets. An important objective of this chapter is to study various operations with fuzzy sets as generalizations of the corresponding operations with ordinary, crisp sets.

Definition

The union and the intersection of two fuzzy sets u, $v: X \to [0, 1]$ are defined by

$$(u \vee v)(x) = \max[u(x), v(x)]$$
$$(u \wedge v)(x) = \min[u(x), v(x)]$$

The complement of a fuzzy set $u: X \to [0, 1]$ is defined by

$$\bar{u}(x) = 1 - u(x)$$

These operations have many properties that are similar to the properties of the classical operations:

(1) $u \vee (v \vee w) = (u \vee v) \vee w$ (associativity)
(2) $u \wedge (v \wedge w) = (u \wedge v) \wedge w$
(3) $u \vee 0 = u, u \wedge 0 = 0$
(4) $u \vee 1 = 1, u \wedge 1 = u$
(5) $u \vee v = v \vee u$
(6) $u \wedge v = v \wedge u$ (commutativity)
(7) $u \wedge (v \vee w) = (u \wedge v) \vee (u \wedge w)$
(8) $u \vee (v \wedge w) = (u \vee v) \wedge (u \vee w)$ (distributivity)
(9) $u \wedge (u \vee v) = u, u \vee (u \wedge v) = u$ (absorption)
(10) $\overline{u \vee v} = \bar{u} \wedge \bar{v}, \overline{u \wedge v} = \bar{u} \vee \bar{v}$ (De Morgan's laws)
(11) $\bar{\bar{u}} = u$ (Double negation law)
(12) $\bar{0} = 1, \bar{1} = 0$

Here $u, v, w \in F(X)$, and 0 and 1 are fuzzy sets:

$$0(x) = 0 \quad 1(x) = 1$$

There is an essential difference between $F(X)$ and $P(X)$: $P(X)$ is a Boolean algebra in the sense that the complement satisfies

$$A \cap \bar{A} = \emptyset \quad A \cup \bar{A} = X$$

$F(X)$ has a more general structure, because

$$u \wedge \bar{u} \neq 0 \quad u \vee \bar{u} \neq 1$$

The only information that we have here is

$$u \wedge \bar{u} \leq \tfrac{1}{2} \quad u \vee \bar{u} \geq \tfrac{1}{2}$$

Because $F(X)$ does not form a Boolean algebra, the concept of fuzzy set is a nontrivial generalization of the concept of ordinary set.

Examples

1. Consider the statement "numbers much larger than 10." This statement about real numbers is inexact; it can be described by a fuzzy subset of the

real line R. Therefore, one can define a membership function $u: R \to [0, 1]$ such that the larger the difference $x - 10$, the closer $u(x)$ is to 1. It is clear that u should at least satisfy the requirements

1. $u(x) = 0$ if $x < 10$.
2. u is nondecreasing.
3. $\lim_{x \to \infty} u(x) = 1$.

A possible formula for u is

$$u(x) = \begin{cases} \max[1 - (x - 10)^{-2}, 0] & \text{if } x > 10 \\ 0 & \text{otherwise} \end{cases}$$

2. Consider the statement "all points in n-space that are very close to the origin." An appropriate membership function can be $u: R^n \to [0, 1]$, defined by

$$u(x) = \exp(-\|x\|^2)$$

where $\exp a = e^a$ and $\|x\| = (\Sigma_{i=1}^{n} x_i^2)^{1/2}$ denotes the Euclidean norm in R^n.

The inclusion of fuzzy subsets is defined by pointwise inequality between the membership functions. Thus, if $u, v \in F(X)$, we have $u \leq v$ if $u(x) \leq v(x)$ for every $x \in X$.

Clearly $0 \leq u \leq 1$ for every $u \in F(X)$ [this relation generalizes the obvious inclusions $\emptyset \subseteq A \subseteq X$ for $A \in P(X)$]. At this point, let us go more deeply into the structure of $F(X)$. Much of this structure is actually inherited from the structure of the unit interval $[0, 1]$ with the operations

1. $a \vee b = \max(a, b)$.
2. $a \wedge b = \min(a, b)$.
3. $\bar{a} = 1 - a$.

So far, because of properties 1–12, the structure $(F(X), \vee, \wedge)$ was viewed as a distributive lattice. Actually, $F(X)$ is a completely distributive lattice, since it is possible to define the union and the intersection of any family of fuzzy sets in such a way that the distributive laws hold. More specifically, let $(u_i)_{i \in I}$ be a family with $u_i \in F(X)$. Then $\bigvee_{i \in I} u_i$ and $\bigwedge_{i \in I} u_i$ are defined by

$$\left(\bigvee_{i \in I} u_i\right)(x) = \sup_{i \in I} u_i(x)$$

$$\left(\bigwedge_{i \in I} u_i\right)(x) = \inf_{i \in I} u_i(x)$$

The distributivity laws are

(13) $v \wedge (\bigvee_{i \in I} u_i) = \bigvee_{i \in I} (v \wedge u_i)$
(14) $v \vee (\bigwedge_{i \in I} u_i) = \bigwedge_{i \in I} (v \vee u_i)$

for every $u_i, v \in F(X)$. It is quite easy to prove these formulae, so they are left to the reader.

Image and Inverse Image

The next important concepts are image and inverse image of a fuzzy set. Let X and Y be two sets and $f: X \to Y$ a function that associates with each element $x \in X$ a unique element $y = f(x) \in Y$. The classical set-theoretic concepts of image and inverse image are defined as follows:

$$A \in P(X) \Rightarrow f(A) = \{y \in Y \mid \exists x \in A, y = f(x)\}$$
$$B \in P(Y) \Rightarrow f^{-1}(B) = \{x \in X \mid f(x) \in B\}$$

Therefore the function f induces two other functions:

$$f: P(X) \to P(Y) \quad f^{-1}: P(Y) \to P(X)$$

It is possible to generate such functions between fuzzy subsets. Let $u \in F(X)$. The image of u under f is defined as $f(u) \in F(Y)$ such that

$$(f(u))(y) = \begin{cases} \sup_{f(x)=y} u(x) & \text{if } y \in f(X) \\ 0 & \text{otherwise} \end{cases}$$

If $v \in F(Y)$, the inverse image of v under f is $f^{-1}(v) \in F(X)$, defined by

$$(f^{-1}(v))(x) = v(f(x))$$

These two relations define two functions

$$f: F(X) \to F(Y) \quad f^{-1}: F(Y) \to F(X)$$

Some simple properties of these definitions (which generalize classical properties) are

(a) $f(\bigvee_{i \in I} u_i) = \bigvee_{i \in I} f(u_i)$
(b) $f(\bigwedge_{i \in I} u_i) \leq \bigwedge_{i \in I} f(u_i)$
(c) $f^{-1}(\bigvee_{i \in I} v_i) = \bigvee_{i \in I} f^{-1}(v_i)$
(d) $f^{-1}(\bigwedge_{i \in I} v_i) = \bigwedge_{i \in I} f^{-1}(v_i)$
(e) $f(f^{-1}(v)) \leq v, f^{-1}(f(u)) \geq u$

where $u, u_i \in F(X)$, $v, v_i \in F(Y)$.

Our next important concept is that of a fuzzy relation. This concept can be thought of as a generalization of classical relations and, in particular, as a generalization of the concept of a function. Let X and Y be two sets.

Definition

A fuzzy relation from X to Y is a fuzzy subset of $X \times Y$—that is, a function $R: X \times Y \to [0, 1]$.

One easy way of generating fuzzy relations is by the product of fuzzy sets. Let $u \in F(X)$ and $v \in F(Y)$. A fuzzy relation R can be defined by

$$R = u \times v: X \times Y \to [0, 1]$$

$$(u \times v)(x, y) = u(x) \wedge v(y), \quad x \in X, y \in Y$$

If R is a fuzzy relation from X to Y, and if S is a fuzzy relation from Y to Z, it is possible to define a fuzzy relation $S \circ R$ from X to Z. The composition of fuzzy relations is defined by

$$(S \circ R)(x, z) = \bigvee_{y \in Y} [R(x, y) \wedge S(y, z)]$$

An important class of fuzzy relations is the class of similarity relations. They generalize the concept of an equivalence relation.

Definition

A fuzzy relation $R: X \times X \to [0, 1]$ is called a *similarity relation* if it has the following properties:

(1) $R(x, x) = 1$ for every $x \in X$ (reflexivity).
(2) $R(x, y) = R(y, x)$ for every $x, y \in X$ (symmetry).
(3) $\bigvee_{z \in X} [R(x, z) \wedge R(z, y)] \leq R(x, y)$ for every $x, y \in X$ (transitivity).

Clearly, in terms of composition, property (3) can be written as $R \circ R \subseteq R$.

ALGEBRAIC THEORIES OF FUZZY SETS

In this section we will study fuzzy sets and functions from a more general, algebraic perspective, using the abstract concepts of category and functor. These concepts will be useful for clarifying the role of the ordering in comparing the membership degrees. From this perspective new generalizations will emerge, such as *L*-sets and *C*-sets.

A *category* is a complex concept consisting of a class of *objects* denoted by $|C|$ and, for any two objects $A, B \in |C|$, a set of *morphisms* from A to B. The latter set is denoted by $C(A, B)$.

If $f \in C(A, B)$ we write $f: A \to B$, or simply use the *diagram* $A \xrightarrow{f} B$. The objects in $|C|$ are not necessarily sets. They can be categories (as we will explain later), the points in a fixed set, or sets with some structure. Therefore, a morphism $A \to B$ is not necessarily a function between sets.

However, it turns out that the simplest example of a category is the category of sets, denoted by Set. Its objects are indeed sets (denoted by A, B, ...), and for each $A, B \in |\text{Set}|$ the morphisms are all functions $f: A \to B$. It is important that the reader keep this example in mind throughout this discussion.

To complete our definition of a category, we recall an important feature of Set: functions between suitably chosen sets can be composed. So we assume, in C, that whenever $f \in C(A, B)$ and $g \in C(B, D)$ we can find a unique morphism $h = g \circ f \in C(A, D)$. In terms of diagrams, $A \to B$ and $B \to D$ mean $A \to D$. The new morphism $g \circ f$ is called the composition of f and g. To have an operational definition and to eliminate trivial situations, we impose the following axioms:

(C1) The composition of morphisms is associative:

$$h \circ (g \circ f) = (h \circ g) \circ f$$

(C2) For each object $A \in |C|$ there exists a morphism $1_A \in C(A, A)$, called the unit morphism, such that for $f \in C(A, B)$ and $g \in C(B, A)$

$$f \circ 1_A = f \quad 1_A \circ g = g$$

(C3) If $C(A, B) \cap C(A', B') \neq \emptyset$, then $A = A'$, $B = B'$.

Again, it is important to note the example of the category Set, where all properties are obviously satisfied. Other examples of categories are the category of groups (where the objects are groups and the morphisms are group homomorphisms) and the category of topological spaces (where the objects are topological spaces and the morphisms are continuous mappings).

The main advantage of the concept of a category is that it provides a common language for various mathematical objects such as sets, topological spaces, algebraic structures, and so on. We will see shortly that fuzzy sets can be considered in this framework.

It was already clear that the ordering structure of $[0, 1]$ and its properties plays an important role in the definition of a fuzzy set. Actually the concept of fuzzy set can be generalized to that of an L-fuzzy set, where L is an arbitrary lattice (instead of the interval $[0, 1]$). A lattice L is a partially ordered set, that is, a set L with an ordering $\leq$ that is reflexive, antisymmetric, and transitive, such that for any two elements $a, b \in L$ there exist two new elements $(a \wedge b), (a \vee b) \in L$, called minimum and maximum, respectively, where

(i) $a \wedge b \leq a, b \leq a \vee b$.
(ii) If $c \leq a, b \leq d$ for $c, d \in L$, then $c \leq a \wedge b \leq a \vee b \leq d$.

The interval $[0, 1]$ is a lattice under the usual ordering and the usual meaning of $\wedge$ and $\vee$ as min and max. The power set $P(M)$ of a set M is a lattice under set inclusion (the latter is a partial ordering and not a total ordering because there are subsets S, T of M such that $S \not\subset T$ and $T \not\subset S$), with $\wedge$ and $\vee$ standing for intersection and union, respectively.

Let X be a set and L a lattice.

Definition

An L-fuzzy subset of X (or simply, L-set) is a function $u: X \to L$. The collection of all L-fuzzy subsets of X will be denoted by

$$F_L(X) = \{u \mid u: X \to L\} = L^X$$

It should be quite clear at this stage that the structure of L is inherited by $F_L(X)$. In particular, $F_L(X)$ becomes a lattice under the ordering

$$u \leq v \Leftrightarrow u(x) \leq v(x) \quad \text{for all } x \in X$$

This is the inclusion of L-sets. The lattice operations $\wedge$ and $\vee$ correspond to the intersection and the union of L-sets:

$$\begin{aligned}(u \wedge v)(x) &= u(x) \wedge v(x) \\ (u \vee v)(x) &= u(x) \vee v(x)\end{aligned} \quad x \in X$$

In general, we do not have a complement in $F_L(X)$ because we did not assume the existence of a complement in L. We do not have the concepts of "empty set" and "total set" in $F_L(X)$ unless we assume the existence in L of a smallest element 0 and a greatest element 1.

Actually, a lattice L (and, in fact, every partially ordered set) can be thought of as a category $\hat{L}$ in the following way: The objects $|\hat{L}|$ are the elements $a, b, c, \ldots \in L$. If $a, b \in |\hat{L}|$, it is assumed that a unique morphism exists from a into b, if $a \leq b$. If $a \not\leq b$, then $\hat{L}(a, b)$ is empty.

How is the composition of morphisms defined? If $f \in \hat{L}(a, b)$ and $g \in \hat{L}(b, c)$, then it means that $a \leq b, b \leq c$. By the transitivity of $\leq$, it follows that $a \leq c$, such that there is a unique morphism, denoted by $g \circ f$ in $\hat{L}(a, c)$.

The *category of L-fuzzy sets* can now be defined as follows: Its objects are pairs (X, u) with X a set and $u: X \to L$ an L-fuzzy subset of X. Amorphism from (X, u) into (Y, v) is simply a function $f: X \to Y$ such that $v \circ f \geq u$.

In terms of the inverse image of a fuzzy set, the last inequality can be written as $u \leq f^{-1}(v)$.

The composition of morphisms is performed in an obvious way. If

$$(X, u) \xrightarrow{f} (Y, v) \text{ and } (Y, v) \xrightarrow{g} (Z, w),$$

then $g \circ f$ is simply the composition of functions f and g. Since $v \circ f \geq v$ and $w \circ g \geq v$, it is easy to show that $w \circ (g \circ f) \geq u$, so $g \circ f$ is well defined.

The identity morphism

$$(X, u) \xrightarrow{1_X} (X, u)$$

is actually

$$1_X: X \to X \qquad 1_X(x) = x$$

which makes sense because $u \circ 1_X = u \geq u$.

It is a simple exercise to show that the concept defined above satisfies the axioms of a category. From now on, Set(L) will denote this category of fuzzy sets.

The category of fuzzy sets can be defined differently: for example, the category Set$_f(L)$ whose objects are pairs (X, u), with $u \in F_L(X)$ as before, and whose morphisms $(X, u) \to (Y, v)$ are fuzzy relations $R: X \times Y \to L$ such that $R(x, y) \leq u(x) \wedge v(y)$ for every $x \in X$, $y \in Y$. The composition of morphisms is the composition of fuzzy relations.

If C and C' are two categories, one would like to be able to associate with an object of C an object of C', and with a morphism in C a morphism in C'. This can be accomplished through the concept of a *functor*.

Definition

A functor Fun: $C \to C'$ is an assignment of an object Fun $A \in |C'|$ to each object $A \in |C|$ and of a morphism Fun $f \in C'($Fun A, Fun $B)$ to each morphism $f \in C(A, B)$ such that the following properties are satisfied:

(F1) Fun$(g \circ f)$ = Fun $g \circ$ Fun f.
(F2) Fun$(1_A) = 1_{\text{Fun}A}$.

As a simple example of functor consider Fun: Set$(L) \to$ Set, where Fun$(X, u) = X$ and Fun $f = f$ whenever

$$(X, u) \xrightarrow{f} (Y, v)$$

is a morphism in Set(L). Such a functor "forgets" the structure of Set(L).

It is important to note that there are various constructions in a category C, such as those leading to products, coproducts, direct and inverse limits, pullback, and so on. These constructions use the objects and morphisms in C as well as some properties defined by diagrams in C, which are directed graphs having objects as vertices and morphisms as edges. Via some of these constructions done in the category Set(L), it is possible to justify some of the concepts in fuzzy set theory, such as the subset of a fuzzy set, intersection, union, and product of fuzzy sets.

So far we have defined fuzzy sets, L-fuzzy sets, and various operations. We have also described the concept of fuzzy set from a categorial point of view. Actually, it is possible to take this generalization further by defining the concept of a C-set, where C is a category. This concept extends that of an L-set and, in particular, that of a fuzzy set in a nontrivial way.

Let C be a category whose class of objects $|C|$ is actually a set. Such a category is called a small category.

Definition

A C-subset of a set X (or a C-set for short) is a function $u\colon X \to |C|$.

As we have described before, a lattice L can be thought of as a category $\hat{L}$. In this case, $C = \hat{L}$, a C-set becomes precisely an L-set; for $L = [0, 1]$ we recapture the concept of a fuzzy set.

What could be the purpose of such a generalization? The concept of "degree of membership" is a complex one. If we wish to model it via the elements of a lattice L (not necessarily totally ordered), it makes sense to assume a more complicated structure. In particular, in what is called fuzzy logic, the degree of membership can be itself a fuzzy set. If the referential (or total) set is not assumed to be the same, we get a Set(L)-set. So, informally speaking, the concept of C-set has to do with assuming the possibility of a structure for the degree of membership of each point.

The collection of all C-subsets of X will be denoted by $F_C(X)$:

$$F_C(X) = \{u \mid u\colon X \to |C|\}$$

Let us now investigate the structure of $F_C(X)$. Clearly, we need not expect that $F_C(X)$ will form a lattice, because as we have seen, the structure is inherited from that of C.

If for each point $x \in X$ we consider the object $u(x) \in |C|$ as a model of the membership degree, then various points $x, y, \ldots \in X$ can be composed via the morphisms (in C) between $u(x), u(y), \ldots$. More precisely, $F_C(X)$ forms a category defined in the following way. The objects of $F_C(X)$ are C-sets $u\colon X \to |C|$. A morphism $m\colon u \to v$, where u, v are C-sets, is a family

$$m = (m_x)_{x \in X} \qquad m_x \in C\big(u(x), v(x)\big) \quad \text{for each } x \in X$$

The composition of morphisms is done pointwise.

If we want to summarize, we can say that the concept of a fuzzy set, where the membership degree was evaluated by numbers in the unit interval, can be extended to that of an L-fuzzy set and carried further to the concept of a C-set, where it is possible to consider morphisms between degrees of membership.

Although this elegant development has generated much research over the past two decades, we still did not answer the vital question, "How can one build a fuzzy set"? Put differently, the question is, "Can one represent fuzzy sets in a different, equivalent, practical way—that is, useful in practical applications"? An answer to this question will be given in the next section, where we discuss the representation theory for fuzzy systems.

REPRESENTATION OF FUZZY SETS

Consider a fuzzy set $u: X \to [0, 1]$ and let α be a given number between 0 and 1. Then clearly the set of elements in X whose membership degree is at least α constitutes an approximation of the fuzzy set u. The set

$$L_\alpha u = \{x \in X \mid u(x) \geq \alpha\}$$

is called the α-*level set* of u.

Other level sets can be defined, such as

$$\{x \in X \mid u(x) > \alpha\}$$
$$\{x \in X \mid u(x) \leq \alpha\}$$
$$\{x \in X \mid u(x) < \alpha\}$$

but, for the purpose of our theory, $L_\alpha u$ is the most meaningful, since it consists of all points whose membership degrees exceed the "threshold" α.

Consider, as example, the statement "x is a small number." This is a fuzzy statement that can be described by a function $u: R \to [0, 1]$. However, sometimes such a statement is equated to the statement $|x| \leq \epsilon$, for some suitable positive number ϵ. It is clear that $\{x \mid |x| \leq \epsilon\}$ is a level set of the fuzzy set u describing all real numbers that are small. In this example the fuzzy set was approximated by a single level set.

However, such an approximation is not correct, and we definitely have to consider all level sets instead of only one.

Next consider the concept of the limit of a sequence, a familiar concept from calculus. The expression $\lim_{n \to \infty} a_n = a$ means that

$$(*) \forall \epsilon > 0, \exists n_\epsilon \text{ with } n > n_\epsilon \Rightarrow |a_n - a| < \epsilon$$

Sometimes this definition is stated as follows: "a_n can be made as close to a as we want, provided that n is large enough." Clearly this is a fuzzy statement that is made precise by $(*)$. The statement $(*)$, on the other hand, is a statement about levels of fuzzy sets.

Assume that $u: X \to [0, 1]$ is a fuzzy set and that $L_\alpha u$ denotes the α-level set, for each $0 \leq \alpha \leq 1$. It is easy to show the following properties:

(1) $L_0 u = X$.
(2) $\alpha \leq \beta \Rightarrow L_\alpha u \supseteq L_\beta u$.

Property (2) shows that the level sets $(L_\alpha u)_{0 \leq \alpha \leq 1}$ form a nested family of sets.

The most important question from the practical point of view is the following: Given a family of subsets of X, $(A_\alpha)_{0 \leq \alpha \leq 1}$, does there exist a fuzzy set $u: X \to [0, 1]$ such that $A_\alpha = L_\alpha u$ for each $\alpha \in [0, 1]$?

It is one goal of this book to show that in many constructions involving fuzzy concepts one has to answer this question. To gain some insight on the importance of this question, assume we want to define a concept based on fuzzy sets $v, w, \ldots$. A first step is to approximate these fuzzy sets by their levels $L_\alpha v, L_\alpha w, \ldots$, and to operate on these levels, which are crisp sets, in a known fashion (to add them if X is a vector space, or to integrate them if the fuzzy sets in question are values of a fuzzy random variable). The result of these operations with sets is another set A_α, for $0 \leq \alpha \leq 1$. More specifically, A_α is some function of $L_\alpha v, L_\alpha w, \ldots$. Now we would like to put all these pieces A_α back together; that is, we would like to define a new fuzzy set u such that A_α is an approximation of u at level α. What this means is exactly what the foregoing problem described: to find a fuzzy set u such that $L_\alpha u = A_\alpha$ for each $\alpha \in [0, 1]$.

Clearly to affirmatively answer this question, we must have that the sets $(A_\alpha)_\alpha$ satisfy at least properties (1) and (2).

Unfortunately, (1) and (2) are not enough for the existence of $u: X \to [0, 1]$ with $L_\alpha u = A_\alpha$. In other words, these two conditions do not guarantee the existence of the fuzzy set u with specified level sets (as mentioned too often in the literature). To prove this statement, we will give a simple example.

Let $X = R$ and consider the family of subsets of R given by

$$A_\alpha = \begin{cases} R & \text{if } 0 \leq \alpha < 1 \\ \varnothing & \text{if } \alpha = 1 \end{cases}$$

There is no fuzzy set $u: R \to [0, 1]$ such that $L_\alpha u = A_\alpha$ for $0 \leq \alpha \leq 1$. The reason is that, although properties (1) and (2) are trivially satisfied, if we consider the sequence $1 - 1/n$, whose limit is 1 as $n \to \infty$, we see that $L_{1-1/n} u = R$, $L_1 u = \varnothing$, but we should have $\bigcap_{n=1}^{\infty} L_{1-1/n} u = L_1 u$, which is false.

The extra property needed for the family $(A_\alpha)_\alpha$ is

(3) If $\alpha_1 \leq \alpha_2 \leq \cdots$ and $\lim \alpha_n = \alpha$, then $A_\alpha = \bigcap_{n=1}^{\infty} A_{\alpha_n}$

This is a left-continuity dependence of A_α on α. We are now ready to state the main result of this section, the representation theorem for fuzzy sets.

THEOREM: Let $(A_\alpha)_{0 \leq \alpha \leq 1}$ be a family of subsets of the space X with the following properties:

(1) $A_0 = X$.
(2) $\alpha \leq \beta \Rightarrow A_\alpha \supseteq A_\beta$.
(3) $\alpha_1 \leq \alpha_2 \leq \ldots$,

$$\lim_{n \to \infty} \alpha_n = \alpha \Rightarrow A_\alpha = \bigcap_{n=1}^{\infty} A_{\alpha_n}$$

There exists a unique fuzzy set $u: X \to [0, 1]$ such that $L_\alpha u = A_\alpha$ for each $\alpha \in [0, 1]$.

Proof: Define the fuzzy set u by

$$u(x) = \sup\{\beta \in [0, 1] \mid x \in A_\beta\} \quad x \in X$$

Let us show first that $A_\alpha \subseteq L_\alpha u$. Choose $x \in A_\alpha$. Then

$$\alpha \in \{\beta \mid x \in A_\beta\}$$
$$\alpha \leq \sup\{\beta \mid x \in A_\beta\} = u(x)$$

Since $u(x) \geq \alpha$, this means that $x \in L_\alpha u$. Now let us show the inclusion $L_\alpha u \subseteq A_\alpha$. Take $x \in L_\alpha u$, so $u(x) \geq \alpha$. This means

$$\sup\{\beta \mid x \in A_\beta\} \geq \alpha$$

If we let $\beta_0 = \sup\{\beta \mid x \in A_\beta\}$, then there exists a sequence $\beta_n \in \{\beta \mid x \in A_\beta\}$, $\beta_1 \leq \beta_2 \leq \cdots$, $\lim_{n \to \infty} \beta_n = \beta_0$. Also recall that $\beta_0 \geq \alpha$. From property (3) it follows that $A_{\beta_0} = \bigcap_{n=1}^{\infty} A_{\beta_n}$. Since $x \in A_{\beta_n}$ for every n, we get that $x \in A_{\beta_0}$. Now, since $\beta_0 \geq \alpha$, property (2) implies $A_{\beta_0} \subseteq A_\alpha$. Since we showed that $x \in A_{\beta_0}$, this implies that $x \in A_\alpha$, so the inclusion $L_\alpha u \subseteq A_\alpha$ is proved. Thus $L_\alpha u = A_\alpha$. The uniqueness of the fuzzy set u is clear. Suppose there were two fuzzy sets u and v with

$$L_\alpha u = L_\alpha v = A_\alpha$$

for each $0 \leq \alpha \leq 1$. Fix an $x_0 \in X$. Obviously $x_0 \in L_{u(x_0)} u$, and since $L_\alpha u = L_\alpha v$ it follows that $x_0 \in L_{u(x_0)} v$. Thus $v(x_0) \geq u(x_0)$. Similarly we show that $u(x_0) \geq v(x_0)$; therefore it follows that $u(x_0) = v(x_0)$. So $u = v$, and the fuzzy set u with $L_\alpha u = A_\alpha$ for each α is unique. The proof is now completed.

Example 1: Addition of Fuzzy Sets

Let $X = R^n$ be Euclidean, n-dimensional space, and consider two fuzzy subsets of R^n:

$$u, v: R^n \to [0, 1]$$

We define $u + v$ by reducing the addition to adding two ordinary sets M, N $\subseteq R^n$:

$$M + N = \{x + y \mid x \in M, y \in M\}$$

This pointwise addition is called the Minkowski addition of sets.

Suppose we want to define $u + v$ so that

$$L_\alpha(u + v) = L_\alpha u + L_\alpha v$$

for any $\alpha \in [0, 1]$. If we denote the right term of this equality by A_α, it is easy to show that the family of sets $(A_\alpha)_\alpha$ satisfies (1)–(3) in the representation theorem. Properties (1) and (2) are obvious; it is the verification of (3) that is important whenever this representation is used. All these verifications are left to the reader. It suffices to say that the representation theorem can be applied and the family $(A_\alpha)_\alpha$ generates a fuzzy set w such that

$$L_\alpha w = A_\alpha = L_\alpha u + L_\alpha v$$

By definition, $w = u + v$.

Example 2: Multiplication of a Fuzzy Set by a Scalar

We start again with a crisp set $M \subseteq R^n$. If $s \in R$ is a scalar, then

$$sM = \{sx \mid x \in M\}$$

describes the multiplication of a set by a scalar.

If $u: R^n \to [0, 1]$ is a fuzzy set, then we let

$$A_\alpha = sL_\alpha u \quad \text{for } 0 \leq \alpha \leq 1$$

Then we show that $(A_\alpha)_\alpha$ satisfies properties (1)–(3). Finally, we define $v = su$ as the unique fuzzy set provided by the representation theorem, such that

$$L_\alpha(su) = sL_\alpha u$$

Example 3: Fuzzy Arithmetic

Fuzzy numbers are fuzzy subsets of the real line R. Sometimes special properties are assumed for fuzzy numbers, such as convexity or normality. For our example these properties are not essential.

Various operations with fuzzy numbers can be defined first on level sets and then extended by using the representation theorem.

Example 4: Generalized Sets

This example is more technical. The central concept here is that of a topos, which is a category T that has finite limits and a subobject classifier I. The last assumption means that there exists a morphism $1 \to I$ (where 1 is the final object in T) such that every subobject $B \to A$ can be obtained by a pullback diagram

$$\begin{array}{ccc} B & \to & 1 \\ \downarrow & & \downarrow \\ A & \to & I \end{array}$$

It is important to note that the familiar category of sets and functions, Set, is a topos and the subobject classifier is $I = \{0, 1\}$. In this case the pullback diagram means that each subset can be represented by its characteristic (or indicator) function (which is $\{0, 1\}$-valued).

There is a much more profound interpretation of this fact: that of the internalization of logic. Indeed, the logic in Set is two-valued, and the truth value object is exactly $\{0, 1\}$. Trying to view the generalization to fuzzy sets in this perspective, one must search for a topos of fuzzy sets such that the truth value object will be the subobject classifier.

Unfortunately, it can be shown that the category Set(L) (where L is a lattice) is not a topos. However, the category Set [0, 1] can be embedded in a topos.

The ideas of representation and level set can be used to define *generalized sets* as families of sets $(A_\alpha)_{\alpha \in L}$ (where L is an ordered set) satisfying a set of restrictions that will always include

$$\alpha, \beta \in L \quad \alpha \leq \beta \Rightarrow A_\alpha \supseteq A_\beta$$

Example 5: Extension Principle

The extension principle says that a function between two sets, $f: X \to Y$, can be extended to fuzzy subsets $f: F(X) \to F(Y)$. As we have mentioned earlier, this can be done as $f(u) = v$, where

$$v(y) = \begin{cases} \sup\{u(x) \mid x \in X, f(x) = y\} & \text{if } f^{-1}(y) \neq \emptyset \\ 0 & \text{otherwise} \end{cases}$$

If we want to express the extension principle in terms of level sets, we would like a formula of the type

$$f(L_\alpha u) = L_\alpha[f(u)] \quad \alpha \in [0, 1]$$

But, in general, this formula is false, although we always have the inclusion

$$f(L_\alpha u) \subseteq L_\alpha[f(u)]$$

To have equality, the representation theorem should be stated in terms of strong level sets. More exactly, if $u: X \to [0, 1]$ and $\alpha \in [0, 1]$, the α-strong level set of u is

$$S_\alpha u = \{x \in X \mid u(x) > \alpha\}$$

In this context

$$f(S_\alpha u) = S_\alpha[f(u)]$$

READINGS

The major topic of this chapter is the concept of a fuzzy set. The name appeared for the first time in

L. A. Zadeh, 1965, Fuzzy sets, *Information and Control* **8**:338-353. This paper introduces a new algebra of classes concerned with new ways of defining new sets. It focuses on the operations of intersection, union, and complementation, defined by the operators MIN and MAX. No explanation is given for this choice, but arguments for their naturalness can be found in

R. E. Bellman, and M. Giertz, 1973, On the analytic formalism of the theory of fuzzy sets, *Information Sciences* **5**:149-156. The operators MAX and MIN are the only ones that exhibit the following necessary properties:

> The membership degree in a compound fuzzy set depends on the membership degrees in those that form it.
> The operators MAX and MIN are commutative, associative, and distributive.
> MIN(1, 1) = 1 and MAX(0, 0) = 0.

Further arguments, following a categorical analysis of logic can be found in

C. V. Negoita, 1985, *Expert Systems and Fuzzy Systems*, Menlo Park, CA: Benjamin/Cummings.

Starting with the observation that the classical rules of logic are represented by operations on the set $\{0, 1\}$, one can use the correspondence between a subset and a characteristic function as illustrated by the pullback diagram

$$\begin{array}{ccc} A & \to & \{1\} \\ \downarrow f & & \downarrow \text{true} \\ X & \to & \{0, 1\} \end{array}$$

The subset A is determined by the characteristic function $X \to \{0, 1\}$. A arises by pulling back $\{1\} \to \{0, 1\}$ along f. In other words, the subset A is the inverse image under f of the subset $\{1\}$ of $\{0, 1\}$.

This observation leads to a morphism-only definition of the truth functions. For instance, the intersection is given by the pullback

$$\begin{array}{ccc} & \text{true, true} & \\ \{0\} & \to & \{0, 1\} \times \{0, 1\} \\ \downarrow & & \downarrow \text{ intersection} \\ \{0\} & \to & \{0, 1\} \end{array}$$

To generalize and to pass from sets to fuzzy sets, we must change the subobject classifier $\{0, 1\}$. This can be done if we notice that a fuzzy set can be *represented* as a family of level sets. In this case the classifier is the set of levels, and the truth morphisms can be defined in the same way.

More technical details can be found in

C. V. Negoita, and A. Stefanescu, 1982, On fuzzy optimization, in *Fuzzy Information and Decision Processes*, M. Gupta and E. Sanchez (eds.), Amsterdam: North-Holland.

In this paper the category of fuzzy sets is viewed as a subcategory of a topos internalizing its own logical calculus. Again, the arguments are based on the representation theorem, according to which a fuzzy set is represented as a family of crisp level sets.

The representation theorem for fuzzy sets was given for the first time in

C. V. Negoita and D. A. Ralescu, 1975, Representation theorems for fuzzy concepts, *Kybernetes* **4**:169–174.

This theorem was successfully used in

M. Sasaki, and M. Sugeno, 1983, L-fuzzy category, *Fuzzy Sets and Systems* **11**:43–64.

Interesting extensions were published by

U. Höhle, 1981, Representation theorems for L-fuzzy quantities, *Fuzzy Sets and Systems* **5**:83–107.

A. Achache, 1982, Galois conexions of a fuzzy set, *Fuzzy Sets and Systems* **8**:215–218.

The need for a theorem of representation was reconfirmed recently when rediscovered in

P. J. Borillo, and R. Fuentes, 1984, A short note on representation of *L*-fuzzy sets by Moore's families, *Stochastica* **8**:291–295.

and used to explain

P. Vopenka, and P. Hajek, 1972, *The Theory of Semisets*, Amsterdam: North-Holland.

In presenting the theory of semisets, the authors hope to make some contribution to the task of breaking through the bars of the prison in which mathematicians find themselves. This prison is set theory, and the authors believe that mathematicians will escape from it just as they escaped from the prison of three-dimensional space.

This is a very powerful statement, and the interested reader can find the basic philosophical assumptions in

A. Sochor, 1984, The alternative set theory and its approach to Cantor's set theory, in *Aspects of Vagueness*, H. Skala, S. Termini, and E. Trillas (eds.), Dordrecht: D. Reidel.

Classical set theory arose during the last century with Bolzano, who considered existing sets. For a set to exist, all its elements must exist individually. So an infinite set is infinite in the actual sense. The first aim of Bolzano's set theory is to prove the existence of infinite sets. In the real world it is impossible to exhibit an infinite set. To produce such a set, we have to abandon the real world. Bolzano therefore employs theological considerations. He proves in a charming and ingenious way, the existence of an infinite set in the "Mind of God." However, actual infinity, or more precisely what is meant by actual infinity, is not a phenomenon. Infinite sets are not something we would be able to perceive. It was Bolzano who knew rather well that only God can observe infinite sets. Mathematicians did not reconcile themselves to Bolzano's intuition, and, since they could find no other, tried to forget it. However, the theological position could not be entirely removed from mathematics. When we say that some real number with particular properties exists, we do not mean that we or anybody else is able to find it, but that God sees it. Moreover, we begin to carry out constructions into which infinitely many objects enter. For example, we create the limit to every sequence fulfilling the Bolzano-Cauchy condition.

Mathematics whose world is Cantor's set theory is mathematics that takes place in some special, not perfect, universe. It is not evident that Cantor's set theory is the best possible description of our comprehension of the real world, and it is quite possible that another theory would enable us to grasp even real-world phenomena, which Cantor's set theory cannot describe in a natural way.

Another approach could be through the concept of a class. A class is mostly understood as described by a property. This means that in the moment of its creation it is not necessary to know all its elements. For example, the property "to be a natural number," and even the class of all natural numbers can be considered before all natural numbers are created. In this way there are sets with subclasses that are not sets. An example is the class of "good men" with objects about which we shall dispute whether they have the property in question.

Sochor mentions that there are more branches of mathematics endeavoring to grasp real situations of the described type: for instance, fuzzy sets and Parik's feasibility or intensional logic. But he does not see a link between fuzzy sets and semisets. This link was noticed, however, in

V. Novak, 1984, Fuzzy sets. The approximation of semisets, *Fuzzy Sets and Systems* **14**:259–272.

Chapter 4

FUZZY RANDOM VARIABLES

The concept of a random variable, so essential in probability and statistical analysis, will be generalized in this chapter. If Ω is the space of all outcomes of a random experiment, (such as flipping a coin, rolling a die, measuring the pressure in an engine), then a function from Ω into the set R of real numbers is called a *random variable*. A function from Ω into n-dimensional Euclidean space R^n is called a *random vector*. Obviously, a random vector is an n-tuple of random variables.

The next step in the generalization is to consider a set varying at random. More precisely, a function from Ω into $\mathcal{P}_0(R^n)$ (the nonempty subsets of R^n) is called a *random set*.

In these definitions we deliberately left unmentioned the hypothesis of measurability. Later in this chapter it will become clear that all these functions from Ω into R, R^n, or $\mathcal{P}_0(R^n)$ are assumed to be measurable in a very precise sense.

Perhaps the oldest problem associated with random sets is the famous needle problem of Buffon: Calculate the probability that a needle placed at random on a plane on which are ruled equidistant parallel lines will intersect one of these lines.

A *fuzzy random variable* is a function from the probability space Ω into the collection of all fuzzy subsets of R^n, $F(R^n)$. In a precise sense, such a variable is a model for a combination of randomness and fuzziness.

RANDOM SETS AND THEIR EXPECTED VALUES

Random Sets

Let $K(R^n)$ denote the collection of all nonempty compact (i.e., closed and bounded) subsets of Euclidean space R^n. The space $K(R^n)$ has a linear structure induced by the Minkowski addition and scalar multiplication

$$A + B = \{a + b | a \in A, b \in B\}$$
$$sA = \{sa | a \in A\}$$

for $A, B \in K(R^n)$ and $s \in R$.

These operations have many nice properties, such as commutativity, associativity, and so on. However, under these operations $K(R^n)$ fails to be a vector space (or linear space). The reason is that even though the opposite of a set A makes sense,

$$-A = \{-a | a \in A\}$$

we do not have $A + (-A) = \{0\}$, unless A is a singleton (it contains precisely one element). In other words, although there are similarities between the structure of $(K(R^n), +, \cdot)$ and the structure of R^n itself, the main difference between these spaces is that in $K(R^n)$ the Minkowski addition does not have an inverse.

Despite this fact, there are yet other similarities between $K(R^n)$ (the space of subsets) and R^n (the space of points). Recall that in R^n we have a distance between points

$$d(x, y) = \left[\sum_{i=1}^{n} (x_i - y_i)^2\right]^{1/2} \quad x, y \in R^n$$

The distance

$$d(x, 0) = \left[\sum_{i=1}^{n} x_i^2\right]^{1/2}$$

is usually denoted by $\|x\|$ (the norm or length of the vector x) and is such that

$$d(x, y) = \|x - y\|$$

It is possible to define a distance in $K(R^n)$ between sets. An immediate choice would be to define, for $A, B \in K(R^n)$, the distance between A and B as the shortest distance between points of A and points of B. However, this would not give a genuine distance function. In particular, the distance between A and B could be zero without having $A = B$.

A proper distance is given by the *Hausdorff distance* (or metric)

$$d(A, B) = \max\left\{\sup_{a\in A}\inf_{b\in B}\|a - b\|, \sup_{b\in B}\inf_{a\in A}\|a - b\|\right\}$$

To understand this abstract definition, note that

$$\inf_{b\in B}\|a - b\| = d(a, B)$$

is the shortest distance from point a to set B. So we can write

$$d(A, B) = \max\left\{\sup_{a\in A} d(a, B), \sup_{b\in B} d(b, A)\right\}$$

The maximum (max) in the definition of $d(A, B)$ is taken to make the distance symmetric.

The properties of Hausdorff distance are

(1) $d(A, B) \geq 0$ and $d(A, B) = 0 \Leftrightarrow A = B$.
(2) $d(B, A) = d(A, B)$ (symmetry).
(3) $d(A, B) \leq d(A, C) + d(C, B)$ (triangle inequality).

We mention that the space $K(R^n)$ with the Hausdorff distance is complete; that is, every Cauchy sequence of sets must be convergent to a set in $K(R^n)$.

So far we have seen that $K(R^n)$ has a linear structure and a distance (it is said to form a metric space). Actually, something equivalent to a norm can be defined in $K(R^n)$ by setting

$$\|A\| = d(A, \{0\}) = \sup_{a\in A}\|a\| \quad A \in K(R^n)$$

which follows immediately from the definition of the Hausdorff distance. In other words, the "length" of a compact set A is measured as the largest distance from the points of A to the origin. It is quite clear why we have to consider compact sets. If the set A were unbounded, then $\|A\|$ could be infinite. As it is, the norm has the following properties:

(a) $\|A\| \geq 0$, $\|A\| = 0 \Leftrightarrow A = \{0\}$.
(b) $\|sA\| = |s|\,\|A\|$, $s \in R$.
(c) $\|A + B\| \leq \|A\| + \|B\|$ (triangle inequality).

The space $K(R^n)$ will be the space of values for our random elements. Occasionally, we will consider a subspace of $K(R^n)$, the space of all compact, nonempty sets that are also convex. The latter will be denoted by $K_c(R^n)$.

Recall that a subset A of R^n is called convex if, for any $a, b \in A$ and $s \in [0, 1]$, we also have $sa + (1 - s)b \in A$. This basically means that the line segment $[a, b]$ is included in A whenever $a, b \in A$.

Obviously, $K_c(R^n) \subseteq K(R^n)$. Moreover, $K_c(R^n)$ is a *closed* subset of $K(R^n)$ in the following sense: If $A_k \in K_c(R^n)$ is a sequence such that A_k converges to a set A in the sense of the Hausdorff distance (i.e., $\lim_{k \to \infty} d(A_k, A) = 0$), then $A \in K_c(R^n)$.

The other piece of information necessary for the definition of a random set is a *probability space*. This is a triple $(\Omega, \mathcal{A}, \mathcal{P})$, where Ω is a set (totality of outcomes of a random experiment), $\mathcal{A}$ is a σ-algebra of subsets of Ω (i.e., $0, \Omega \in \mathcal{A}$; if $A \in \mathcal{A}$, then $\Omega \setminus A \in \mathcal{A}$; and if $A_k \in \mathcal{A}$ for $k = 1, 2, \ldots$, then $\bigcup_{k=1}^{\infty} A_k \in \mathcal{A}$), and $\mathcal{P}$ is a probability measure on $\mathcal{A}$.

The σ-algebra $\mathcal{A}$ contains all (measurable) events related to the random experiment with outcomes in Ω. The probability measure $\mathcal{P}$ is a set function $\mathcal{P}: \mathcal{A} \to [0, 1]$ such that

(i) $\mathcal{P}(\emptyset) = 0$.
(ii) $\mathcal{P}(\bigcup_{k=1}^{\infty} A_k) = \sum_{k=1}^{\infty} \mathcal{P}(A_k)$ if $A_k \in \mathcal{A}$ and
 $A_k \cap A_l = \emptyset$ for $k \neq l$ (i.e., A_k are mutually disjoint).

Roughly speaking, a *random set* is a function

$$X: \Omega \to K(R^n)$$

that associates a compact subset of R^n to each outcome of some random experiment. We can speak about the values of a random set only in probabilistic terms (e.g., the probabilities that X will intersect a given set, or the probability that X will be included in a given set, and so on). For such probabilities to make sense, the events in question should belong to $\mathcal{A}$.

This leads us to the hypothesis of measurability. This concept can be approached in several ways (all equivalent in this context). For our purposes the simplest way, perhaps, is to talk about the *graph* of a random set:

$$G_X = \{(\omega, x) \in \Omega \times R^n \mid x \in X(\omega)\}$$

The function X is called *measurable* if G_X belongs to the σ-algebra. $\mathcal{A} \times \mathcal{B}_{R^n}$ (generated by Cartesian products of events in $\mathcal{A}$ and open subsets of R^n). The graph G_X is indeed a subset of the product space $\Omega \times R^n$.

So the formal definition is

Definition

A *random (compact) set* is a measurable function $X: \Omega \to K(R^n)$.

It is possible to talk about more general concepts such as random closed set or random set in general, or even to consider spaces more general than R^n.

Example 1

The simplest example of a random set is a constant random set $X(\omega) = M$, where $M \in K(R^n)$.

Example 2

Another very simple random set is obtained as follows: Let $\xi: \Omega \to R^n$ be a random vector. Then $X: \Omega \to K(R^n)$, defined by $X = \{\xi\}$, is a random set.

This example shows that random sets generalize random vectors and random variables (obtained if $n = 1$).

Example 3

A nontrivial example of a random set is obtained by adding a constant set to a random vector $X = M + \{\xi\}$, where $M \in K(R^n)$ and $\xi: \Omega \to R^n$ is a random vector. This construction is used for confidence intervals and confidence regions in statistics, a topic to be discussed in detail later.

Example 4

Let $\eta: \Omega \to R$ be a random variable and $M \in K(R^n)$ a fixed set. Define $X = \eta M$. Some confidence intervals in statistics are of this form.

Example 5

Let $A \in \mathcal{C}$ be a fixed event and $M, N \in K(R^n)$ two fixed compact sets. Define a random set X as follows:

$$X(\omega) = \begin{cases} M & \text{if } \omega \in A \\ N & \text{if } \omega \in \Omega \setminus A \end{cases}$$

Such a random set is an extension of a random variable assuming only two values.

More generally, we can define

$$X(\omega) = \begin{cases} M_1 & \text{if } \omega \in A_1 \\ M_2 & \text{if } \omega \in A_2 \\ \ldots \\ M_k & \text{if } \omega \in A_k \end{cases}$$

where $M_1, M_2, \ldots, M_k \in K(R^n)$, $A_1, A_2, \ldots, A_k \in \mathcal{C}$, and $\bigcup_{j=1}^{k} A_j = \Omega$, $A_i \cap A_j = \emptyset$ for $i \neq j$.

Such a random set (assuming finitely many values) can also be written as

$$X = \sum_{j=1}^{k} M_j \chi_{A_j}$$

where χ_{A_j} is the $\{0, 1\}$-valued characteristic function of the event A_j in $\mathcal{C}$.

Expected Value

Our next important problem is that of defining the *expected value of a random set*. This is a sort of average or mean value of the values taken by the random set. Since these values are themselves sets, it is clear that the expected value should be a set rather than a number (or a vector in R^n).

One way of defining the expected value is through the Aumann integral, as described below.

Let $X: \Omega \to K(R^n)$ be a random set. A *selector* of X is a point-valued random vector $\varphi: \Omega \to R^n$ such that $\varphi(\omega) \in X(\omega)$ for each $\omega \in \Omega$. In other words, φ "selects" a point in X in a measurable way. Whether there exist such measurable selections is a nontrivial problem. For our purposes it suffices to say that the answer is affirmative, precisely because we assumed our random set X to be measurable.

There are many selectors of X. We only consider those which have a well-defined expected value $E\varphi$.

In general, there may be no such selectors unless an additional hypothesis is imposed on X. This is exactly the hypothesis concerning finiteness of the first moment, $E\|X\| < \infty$. We will come back to these assumptions later.

The procedure in defining the expected value EX of X is as follows: Consider all selectors of X with finite expectation $E\varphi$; then form the set of all these expectations $E\varphi$. In passing, we note that selectors whose expectation makes sense are said to belong to the space $L^1(\Omega, \mathcal{C}, \mathcal{P})$.

Definition

Let $X: \Omega \to K(R^n)$ be a random set such that $E\|X\| < \infty$. The *expected value* EX of X is

$$EX = \{E\varphi \mid \varphi \text{ is an } L^1(\Omega, \mathcal{C}, \mathcal{P}) \text{ selector of } X\}$$

Although the definition of EX is quite natural, from the practical point of view it would be hard to calculate expected values of random sets by considering all the selectors.

An equivalent way of defining the expectation is by the Debreu integral.

To avoid mathematical complications beyond our scope, we will assume random sets $X: \Omega \to K_c(R^n)$—that is, sets having compact and *convex* values.

Assume first that X takes on two values:

$$X(\omega) = \begin{cases} M & \text{if } \omega \in A \\ N & \text{if } \omega \in \Omega \setminus A \end{cases}$$

We then define

$$EX = P(A)M + [1 - P(A)]N$$

According to the operations with sets (Minkowski addition and scalar multiplication), this definition makes sense.

We say that X is simple if it only assumes finitely many values:

$$X(\omega) = \begin{cases} M_1 & \text{if } \omega \in A_1 \\ M_2 & \text{if } \omega \in A_2 \\ \ldots \\ M_k & \text{if } \omega \in A_k \end{cases}$$

As we have mentioned in Example 5, X can be written as

$$X = \sum_{j=1}^{k} M_j \chi_{A_j}$$

We then define the expectation by

$$EX = \sum_{j=1}^{k} M_j P(A_j)$$

The right side is indeed a weighted average of the values $M_1, M_2, \ldots, M_k$ (sets), the weights being the probabilities $P(A_1), P(A_2), \ldots, P(A_k)$ (scalars).

It is also clear that if X is a simple random set, then $EX \in K_c(R^n)$. If X is an arbitrary (convex-valued) random set with $E\|X\| < \infty$, then EX is obtained from an approximation by expectations EX_j of simple random sets X_j in the following way: There exists a sequence $(X_j)_j$ of simple random sets such that

$$\lim_{j \to \infty} d(X_j, X) = 0$$

where d stands for the Hausdorff distance. The convergence is pointwise. We then define

$$EX = \lim_{j \to \infty} EX_j$$

a limit also in the sense of the Hausdorff distance, because we are dealing with sets.

So EX can be computed from expectations of simple random sets by a limiting process. It is also possible to show that EX does not depend on the particular sequence (X_j) of simple random sets chosen to converge to X.

Note that if $E\|X\| < \infty$ (we say that X is integrable), the two definitions of EX are equivalent.

As we have already observed, the expected value EX of a random set is a set, and it generalizes the expected value of an ordinary random variable or random vector. Sometimes EX is written in terms of an integral as

$$EX = \int_\Omega X\, dP$$

Many interesting properties are related to this expected value. Perhaps one of the most important such properties is a theorem of the Lebesgue dominated convergence type. This result extends its classical counterpart. As the name implies, it is a convergence theorem.

To get some insight into this result, note that if $X_j \to X$ it is not true, in general, that $EX_j \to EX$. In other words, we cannot interchange the limit and the integral as in

$$\lim_{j\to\infty} \int_\Omega X_j\, dP = \int_\Omega \left(\lim_{j\to\infty} X_j\right) dP$$

without additional hypotheses. On the other hand, such a property would be very useful for various constructions.

What exactly are the additional hypotheses required? It basically involves requesting the existence of a dominating function with finite expectation, as in the classical case. The result is the following theorem.

THEOREM: If $X_j: \Omega \to K(R^n)$ are random sets with $\lim_{j\to\infty} d(X_j, X) = 0$, and if there exists a scalar function $h: \Omega \to R$, $Eh < \infty$, such that $\|X_j\| \leq h$ for $j = 1, 2, \ldots$, then

$$\lim_{j\to\infty} d(EX_j, EX) = 0$$

Note again that all the convergences involving random sets and their expectations are in the sense of the Hausdorff distance d. The distance $d(M, N)$ measures the closeness between sets M and N, and it extends the usual distance in R^n since

$$d(\{a\}, \{b\}) = \|a - b\|$$

FUZZY RANDOM VARIABLES

Suppose the following game is played: A coin is flipped; if the outcome is tails, the player wins approximately $10, whereas if the outcome is heads, the player loses a lot of money. This is a very simple example of a fuzzy random variable, a combination of randomness and fuzziness. Randomness is involved through the random experiment of flipping a coin. Fuzziness is involved through the inexactness of the payoffs associated to the random outcomes.

A first question that comes to mind is related to the definition of the expected value of the player in this game. Obviously, this value should be a fuzzy set rather than a number or a set.

Another example is measuring the depth of a lake at a randomly chosen location. The possible values of such a measurement could be deep, very deep, approximately 6 feet, between 5 and 6 feet, and so on.

In this section we will explain the concept of a fuzzy random variable and define its expected value. The main tool will be the representation theorem for fuzzy sets.

Intuitively, a fuzzy random variable is a function X from a probability space Ω into the space of fuzzy subsets of R^n. Measurability conditions should be imposed. As was the case for random sets, we cannot really allow the whole space of fuzzy subsets of R^n to be the set of possible values. For one thing, the values should be "bounded" in some sense. We will see that geometric restrictions play an important role in this context.

The first step is to extend the space of compact sets $K(R^n)$ to a space of fuzzy subsets of R^n. To do this, we have to specify the fuzzy sets should be included, the algebraic structure (linear structure), and the method of measuring the distance between fuzzy sets.

A fuzzy subset of R^n is a function

$$u: R^n \to [0, 1]$$

If $0 \leq \alpha \leq 1$, the α-level set of u is

$$L_\alpha u = \{x \in R^n | u(x) \geq \alpha\}$$

The support of u, denoted by supp u, is the closure of the set $\{x \in R^n | u(x) \neq 0\} = \{x \in R^n | u(x) > 0\}$. So

$$\text{supp } u = \text{cl}\{x \in R^n | u(x) \neq 0\}$$

By closure we mean that, together with the points of the set under consideration, we also include all points that can be obtained as limits of sequences of points in the set. For example

$$\text{cl}(1, 2) = [1, 2]$$
$$\text{cl}\{x \in R^n | \|x\| < 1\} = \{x \in R^n | \|x\| \leq 1\}$$

We will consider as our range space the collection $F(R^n)$ of all fuzzy subsets of R^n with the following properties:

(1) supp u is compact (i.e., closed and bounded).
(2) $L_\alpha u$ is closed for each $0 \le \alpha \le 1$.
(3) $L_1 u \ne \emptyset$.

What is the meaning of these hypotheses? Clearly, they imply that all levels of u are compact and nonempty. Moreover, (1) is the required global boundedness. Indeed, fuzzy sets such as "approximately 3.7," "much greater than 10 and slightly less than 100," "in the neighborhood of the origin of the n-space," and so on, are allowed as possible values. Plainly, they belong to our range space.

However, fuzzy sets such as "much greater than 10" will not be allowed as possible values. The boundedness is violated in this case.

It is clear now that $K(R^n) \subseteq F(R^n)$ in the following sense: If K is a compact nonempty set, its characteristic function belongs to $F(R^n)$.

The next step is to define the linear structure in $F(R^n)$—that is, to define addition of fuzzy sets and multiplication of a fuzzy set by a scalar. These operations can be defined by using level sets as before or by the following intrinsic way: Suppose that $u, v \in F(R^n)$; then $u + v$ is the fuzzy set (membership function) given by

$$(u + v)(x) = \sup_{y+z=x} \min[u(y), v(z)] \quad x \in R^n$$

It is easy to be shown that $u + v$ has properties (1)–(3). It is also true that

$$L_\alpha(u + v) = L_\alpha u + L_\alpha v$$

for each $0 \le \alpha \le 1$.

Scalar multiplication su [where $s \in R$ and $u \in F(R^n)$] is defined in the following way:

$$(su)(x) = \begin{cases} u\left(\dfrac{x}{s}\right) & \text{if } s \ne 0 \\ 0 & \text{if } s = 0 \text{ and } x \ne 0 \\ 1 & \text{if } s = 0 \text{ and } x = 0 \end{cases}$$

This can also be written more compactly as

$$(su)(x) = \begin{cases} u\left(\dfrac{x}{s}\right) & \text{if } s \ne 0 \\ \chi_{\{0\}}(x) & \text{if } s = 0 \end{cases}$$

The level condition is satisfied:

$$L_\alpha(su) = sL_\alpha u \quad 0 \leq \alpha \leq 1$$

These operations do extend the corresponding operations with sets. Namely, if $M, N \in K(R^n)$ then

$$\chi_M + \chi_N = \chi_{M+N} \quad s\chi_M = \chi_{sM}$$

On the other hand, the foregoing definitions are the only choice if we want to satisfy the level conditions

$$L_\alpha(u + v) = L_\alpha u + L_\alpha v$$

$$L_\alpha(su) = sL_\alpha u$$

So far we have shown that the space $F(R^n)$ has a linear structure given by the operations $+$ and $\bullet$. Unfortunately (or fortunately) $F(R^n)$ does not become a vector space because [like $K(R^n)$] there is no opposite of a fuzzy set, in the sense of algebra. It is true that one can define $-u$ for $u \in F(R^n)$ by

$$(-u)(x) = u(-x) \quad x \in R^n$$

but $u + (-u) \neq 0$, in general, where 0 in the right side means $\chi_{\{0\}}$ representing the one point set $\{0\}$.

The last step in describing the structure of $F(R^n)$ is to define a metric (distance) that extends the Hausdorff metric. This can be done in different ways. One possibility is to define

$$d_\infty(u, v) = \sup_{0 \leq \alpha \leq 1} d(L_\alpha u, L_\alpha v)$$

In other words, we approximate u and v by their levels $L_\alpha u$ and $L_\alpha v$, we take the Hausdorff distance between levels, and, then we take the largest distance. Obviously, $d_\infty(u, v)$ is a finite quantity because of the restrictions imposed on u and v.

It is easy to prove that d_∞ is a metric: that is,

(a) $d_\infty(u, v) \geq 0$, $d_\infty(u, v) = 0 \Leftrightarrow u = v$.
(b) $d_\infty(u, v) = d_\infty(v, u)$ (symmetry).
(c) $d_\infty(u, v) \leq d_\infty(u, w) + d_\infty(w, v)$ (triangle inequality).

Another way of defining a distance between fuzzy sets is by the formula

$$d_1(u, v) = \int_0^1 d(L_\alpha u, L_\alpha v) \, d\alpha$$

integrating the distance between levels because $d(L_\alpha u, L_\alpha v)$ is integrable. This distance is a genuine metric satisfying properties (a)–(c).

We can even define a more general metric by

$$d_r(u, v) = \left[\int_0^1 d^r(L_\alpha u, L_\alpha v) \, d\alpha \right]^{1/r}$$

for $r \geq 1$ a real number. Then d_1 is obtained as a particular case of $r = 1$.

What makes us prefer one metric over the other? Note that all of them extend the Hausdorff metric in the sense that if $M, N \in K(R^n)$ then

$$d_\infty(\chi_M, \chi_N) = d_1(\chi_M, \chi_N) = d_r(\chi_M, \chi_N) = d(M, N)$$

The answer to this question is intimately related to the problem of separability of the space $F(R^n)$, which will be discussed later in the chapter.

There are many interesting properties of the space $F(R^n)$ endowed with one of the metrics just described. We just mention here that $F(R^n)$ is complete in the metric d_∞, and this means that every Cauchy sequence of fuzzy sets in $F(R^n)$ must converge to a fuzzy set in $F(R^n)$ with respect to d_∞.

We are now ready to define the concept of a fuzzy random variable. Basically, this will be a function

$$X: \Omega \to F(R^n)$$

from a probability space Ω into our space of fuzzy sets.

To make this concept mathematically meaningful, we impose the hypothesis of measurability.

Note that to each function $X: \Omega \to F(R^n)$ we can associate a random set $X_\alpha: \Omega \to K(R^n)$ in the following way:

$$X_\alpha(\omega) = L_\alpha[X(\omega)]$$

This can be done for every $\alpha \in [0, 1]$. In other words, X_α associates to every sample point ω the α-level set $L_\alpha[X(\omega)]$ of the fuzzy set $X(\omega)$.

Naturally, X_α is a genuine random set only if it is measurable. Now we can give the following formal definition.

Definition

A *fuzzy random variable* is a function $X: \Omega \to F(R^n)$ such that X_α is measurable for each $\alpha \in [0, 1]$.

Example 1

Suppose that X assumes only values u and v:

$$X(\omega) = \begin{cases} u & \text{if } \omega \in A \\ v & \text{if } \omega \in \Omega \setminus A \end{cases}$$

Example 2

If X assumes only finitely many values

$$X(\omega) = \begin{cases} u_1 & \text{if } \omega \in A_1 \\ u_2 & \text{if } \omega \in A_2 \\ \ldots \\ u_k & \text{if } \omega \in A_k \end{cases}$$

where $u_1, u_2, \ldots, u_k \in F(R^n)$, we can write

$$X = \sum_{j=1}^{k} u_j \chi_{A_j}$$

Example 3

Suppose that $\xi: \Omega \to R^n$ is an ordinary random vector. Let $u \in F(R^n)$ be a fixed fuzzy set. Then $X = u + \{\xi\}$ is a fuzzy random variable. Roughly speaking, X is obtained from a constant fuzzy set along a random trajectory.

Example 4

Let $\eta: \Omega \to R$ be a real-valued random variable and $u \in F(R^n)$ a constant fuzzy set. Then one can define a fuzzy random variable by $X = \eta u$.

Example 5

It is possible to combine Examples 3 and 4 to get fuzzy random variables $X = \eta u + \{\xi\}$.

EXPECTED VALUE OF A FUZZY RANDOM VARIABLE

The expected value of a fuzzy random variable can be defined in two ways. The first one uses the level random sets X_α. It is more elegant but mathematically more complicated. The second way, easier to apply in practice, starts with simple fuzzy random variables (i.e., assuming finitely many values) and then extends the expected value through a limiting process. We will describe both approaches here.

Let $X: \Omega \to F(R^n)$ be a fuzzy random variable. We would like to define its expected value EX as a *fuzzy set* in $F(R^n)$ with the property

$$L_\alpha(EX) = E(X_\alpha) \quad 0 \leq \alpha \leq 1$$

We need this property because we want to approximate a fuzzy set by its levels. After all, the random set X_α is itself an approximation to the fuzzy random variable X. We will briefly describe this construction here.

Let X be a fuzzy random variable. Consider the level random sets X_α with expectations EX_α. Clearly, $M_\alpha = EX_\alpha \in K(R^n)$. The family of ordinary sets $(M_\alpha)_{0 \leq \alpha \leq 1}$ can be shown to satisfy the hypotheses of the representation theorem. So $(M_\alpha)_\alpha$ generates a unique fuzzy set u given by

$$u(x) = \sup \{\alpha \epsilon [0, 1] \mid x \in M_\alpha\} \qquad x \in R^n$$

It is this fuzzy set u that we denote by EX, and we call it the expected value of the fuzzy random variable X.

Definition

The *expected value* of the fuzzy random variable X is the unique fuzzy set EX

$$L_\alpha(EX) = E(X_\alpha) \qquad 0 \leq \alpha \leq 1$$

The second approach to the definition of the expected value assumes first that

$$X = \sum_{j=1}^{k} u_j \chi_{A_j}$$

is a simple fuzzy random variable, where $u_j \in F(R^n)$, $A_j \in \mathcal{A}$, and $\mathcal{A}$ is the σ-algebra of events on which a probability measure P is defined. Then set

$$EX = \sum_{j=1}^{k} u_j P(A_j)$$

Note that the right side involves addition of fuzzy sets and multiplication of a fuzzy set by a real number. So, obviously, EX is a fuzzy set, as it should be.

For a more general fuzzy random variable, not necessarily simple, take a sequence of simple fuzzy random variables X_j such that $\lim_{j \to \infty} d_\infty(X_j, X) = 0$. Then define

$$EX = \lim_{j \to \infty} EX_j$$

where EX_j was defined previously for simple fuzzy random variables. The limit of fuzzy sets is in the sense of the metric d_∞.

Keep in mind that many mathematical details have been omitted. The expected value EX is also called the integral of X (with respect to the probability measure P). It is denoted by

$$EX = \int_\Omega X \, dP$$

Example 1

Suppose that a coin is flipped. If the outcome is tails (T), the player loses approximately $10; if the outcome is heads (H), the player wins an amount much larger than $100 but not much larger than $1000. The fuzzy random variable in question is

$$X: \{T, H\} \to F(R)$$

$$X(T) = \text{approximately} -10 = u$$

$$X(H) = \text{much larger than 100 but not much larger}$$

$$\text{than } 1000 = v$$

$$u, v: R \to [0, 1]$$

Some intuitively plausible membership functions for these fuzzy sets are

$$u(x) = \left[1 - \frac{(x+10)^2}{4}\right]^+$$

$$v(x) = \left[1 - \frac{(x-630)^2}{380^2}\right]^+$$

where for a real-valued function f, $f^+ = \max(f, 0)$ is the positive part.

It is easy to show that EX is given by

$$(EX)(x) = \sup_{y+z=2x} \min\left\{\left[1 - \frac{(y+10)^2}{4}\right]^+, \left[1 - \frac{(z-630)^2}{380^2}\right]^+\right\}$$

In particular, the support of the fuzzy set EX is included in the interval $[119, 501]$.

Example 2

Consider a discrete fuzzy random variable $X: \Omega \to F(R)$. This means that X assumes values (fuzzy sets) $u_1, u_2, u_3, \ldots$ with respective probabilities $p_1, p_2, p_3, \ldots$. One can show in a rigorous way that

$$EX = \sum_{i=1}^{\infty} p_i u_i$$

Here an infinite sum of fuzzy sets $\Sigma_{j=1}^{\infty} v_j$ is defined by

$$\left(\sum_{j=1}^{\infty} v_j\right)(x) = \sup \inf_{j \geq 1} [v_j(y_j)]$$

where the supremum is taken over all sequences $\{y_1, y_2, \ldots\}$ such that $\Sigma_{j=1}^{\infty} |y_j| < \infty$ and $x = \Sigma_{j=1}^{\infty} y_j$. We will have more to say about such infinite sums (or series) later.

The expected value of the fuzzy random variable we have just defined has many important properties. Note that we have not discussed the problem of integrability. In other words, the expectation EX may not exist unless an additional hypothesis is imposed.

If X is a fuzzy random variable, let supp X be the random set

$$(\operatorname{supp} X)(\omega) = \operatorname{supp} X(\omega)$$

where supp means support of a fuzzy set. The integrability condition is

$$E \| \operatorname{supp} X \| < \infty$$

where $\| \ \|$ denotes the norm of a set. It is only under this finiteness condition that EX is well defined.

The most important property for our purposes is a convergence theorem of Lebesgue dominated convergence type. As for random sets, this result answers the question of the possibility of interchanging a limit and the expected value for fuzzy random variables.

THEOREM: Let X_j, X be fuzzy random variables such that $E\| \operatorname{supp} X_j \| < \infty$, $E \| \operatorname{supp} X \| < \infty$. Assume that $\lim_{j \to \infty} X_j = X$, distance $(X_j(\omega), \{0\}) \leq h(\omega)$ for all $j \geq 1$, and $h: \Omega \to R$ is integrable. Then

$$EX_j \to EX$$

Note: By distance we mean either d_∞ or d_1. The theorem works in each case. Also the convergence is in the corresponding distance; that is,

$$\lim_{j \to \infty} \operatorname{distance}(X_j, X) = 0$$

$$\lim_{j \to \infty} \operatorname{distance}(EX_j, EX) = 0$$

The proof is left to the reader.

Chapter 5

LIMIT THEOREMS FOR FUZZY RANDOM VARIABLES

In statistics limit theorems are crucial. Statisticians dealing with relatively large samples use them to do computations related to confidence intervals or testing hypotheses. The law of large numbers provides the justification for the assumption that frequencies approach (in the limit) the value of the probability itself. One could say that the limit theorems and the whole theory developed out of limit results constitute the soul of probability theory and distinguish this discipline from measure theory.

Our goal in this book is to develop concepts and techniques for statistical inference based on inexact data. Roughly speaking, this means to incorporate fuzziness into a statistical framework and to use both probabilistic and fuzzy information for decision analysis.

Part of the statistical analysis of imprecise data is concerned with inference based upon samples of fuzzy random variables. To make fuzzy random variables useful in statistical problems, we must develop a theory that will include limit theorems (such as the law of large numbers and the central limit theorem) for fuzzy random variables. Since the concept of fuzzy random variable extends the concept of random variable, we must first discuss limit theorems for random sets.

LIMIT THEOREMS FOR RANDOM SETS

An important concept in this and subsequent sections is that of a Banach space. This concept will be useful for our development because, although $K(R^n)$ and $F(R^n)$ are not genuine vector spaces, they can be *embedded* into such spaces and, more precisely, they can be embedded into Banach spaces.

Recall that the Euclidean space R^n has a norm

$$\|x\| = \left(\sum_{i=1}^{n} x_i^2\right)^{1/2}$$

and that R^n is complete with respect to this norm.

Every vector space $\mathfrak{X}$ (which thus has an addition and a scalar multiplication) on which there is defined a norm such that $\mathfrak{X}$ is complete (i.e., every Cauchy sequence must be convergent) is called a *Banach space*.

The reader might think at this at this stage that R^n is a sufficient example of a Banach space. This is not true.

A more adequate example of Banach space is the following: Denote by $C[0, 1]$ the set of all continuous functions $\varphi: [0, 1] \to R$. The addition and scalar multiplication are defined as usual for functions, pointwise. The norm is defined by

$$\|\varphi\| = \sup_{0 \leq t \leq 1} |\varphi(t)|$$

The space $(C[0, 1], \|\ \|)$ is a Banach space.

The fundamental difference between $C[0, 1]$ and R^n is that the former is infinitely dimensional. What this means is that R^n has a finite basis $e_1, e_2, \ldots, e_n$ (the unit vectors along the coordinate axes), and every vector in R^n is a linear combination of them. $C[0, 1]$, on the other hand, contains an infinite collection of linearly independent elements.

Our next goal now is to describe the strong law of large numbers for random sets. The starting point is a sequence $X_1, X_2, \ldots, X_k, \ldots$ of independent and identically distributed (i.i.d.) random sets.

Recall that our random sets take values in $K(R^n)$, which has a linear structure and is endowed with the Hausdorff distance. We say that $K(R^n)$ is a *metric space*.

A subset U of $K(R^n)$ is called *open* if, for every $W \in U$, there exists an $r > 0$ such that $\{Z \in K(R^n)/d(Z, W) < r\} \subseteq U$. In other words, U is open if, together with every point, it contains a suitable ball centered at that point.

Note that U is a superset. Its elements (points) are sets of $K(R^n)$. The random sets $X_1, \ldots, X_k: \Omega \to K(R^n)$ are called *independent* if

$$P\{X_1 \in U_1, X_2 \in U_2, \ldots, X_k \in U_k\} = P\{X_1 \in U_1\} \cdot P\{X_2 \in U_2\} \cdots P\{X_k \in U_k\}$$

for every open $U_1, U_2, \ldots, U_k \subseteq K(R^n)$.

A sequence of random sets $X_1, X_2, \ldots, X_k, \ldots$ is called *independent* if every finite collection in the sequence satisfies the preceding property. Clearly, this concept is a generalization of independence for ordinary random variables.

Random sets $X_1, X_2, \ldots, X_k, \ldots$ are called *identically distributed* if

$$P\{X_1 \in U\} = P\{X_2 \in U\} = \cdots = P\{X_k \in U\} = \cdots$$

for every open set $U \subseteq K(R^n)$.

From now on, when dealing with i.i.d. random sets, we shall mean that the two preceding conditions are satisfied.

The concept of a convex hull of a set needs to be clarified before we proceed further. If W is a set in $K(R^n)$, its *convex hull* co W is the smallest convex set containing W. For example, if a, b are two different points in a plane, then co $\{a, b\}$ is the segment line connecting a and b.

From any random set X we can obtain a convex-valued random set co X by setting

$$(\text{co } X)(\omega) = \text{co}[X(\omega)] \quad \omega \in \Omega$$

We are now ready to state the strong law of large numbers for random sets.

THEOREM: Let $X_1, X_2, \ldots, X_k, \ldots$ be i.i.d. random sets with $E\|X_1\| < \infty$. Then

$$d\left(\frac{X_1 + X_2 + \cdots + X_k}{k}, E(\text{co } X_1)\right) \to 0$$

almost surely, as $k \to \infty$.

Note that the condition $E\|X_1\| < \infty$ is essential. The theorem shows that the average of sets $k^{-1}(X_1 + X_2 + \cdots + X_k)$ converges (in the Hausdorff distance) to the common expected value $E(\text{co } X_1)$ of the convex hull of X_1. In this respect, we say that the average is a convexifying operation. Note that we did not start with random sets with convex values.

Convergence almost surely means pointwise convergence (at every sample point $\omega \in \Omega$) except at points in a set of probability zero.

This theorem is quite deep, although it is nothing more than a generalization of the familiar result that says if we flip a coin many times, say n, and if we get j tails out of n flips, then the frequency (relative number of tails) j/n approaches $\frac{1}{2}$ (the probability of tails) as $n \to \infty$.

Our next main goal in this section is to describe the central limit theorem for random sets. The classical central limit theorem says that if $\xi_1, \xi_2, \ldots, \xi_n, \ldots$ are i.i.d. random variables with a finite variance, then the distribution of

$$\sqrt{n}\left(\frac{1}{n}(\xi_1 + \xi_2 + \cdots + \xi_n) - E\xi_1\right)$$

is approximately normal with mean zero if n is large. This result is used to calculate probabilities about averages of random variables by using the well-known normal probabilities.

The result for random sets is more complicated. For one thing, it is not a result about the difference between an average and the expected value because we are dealing with sets. Also, the limiting distribution is no longer a simple univariate normal distribution.

Some notations need to be defined. Let

$$S^{n-1} = \{x \in R^n / \|x\| = 1\}$$

be the unit sphere in R^n. By $C(S^{n-1})$ we denote all continuous functions $\varphi: S^{n-1} \to R$. The norm in $C(S^{n-1})$ is given by

$$\|\varphi\| = \sup_{x \in S^{n-1}} |\varphi(x)|$$

It can be shown that $C(S^{n-1})$ becomes a Banach space (infinite dimensional).

When can we say that a variable $Z: \Omega \to C(S^{n-1})$, valued in $C(S^{n-1})$, is normal (or Gaussian)? Note that if Z is such a variable (a sort of stochastic process) and if $t \in S^{n-1}$, then one can define $Z(t): \Omega \to R$, an ordinary random variable, by

$$Z(t)(\omega) = Z(\omega)(t) \qquad \omega \in \Omega$$

We say that Z is *Gaussian* if, for every $t_1, t_2, \ldots, t_r \in S^{n-1}$, the random vector $(Z(t_1), Z(t_2), \ldots, Z(t_r))$ has a multivariate normal distribution (in R^r).

Loosely speaking, the process Z is Gaussian if its finitely dimensional distributions are normal. Such a Z is valued in $C(S^{n-1})$, a Banach space. Our random sets take values in $K(R^n)$, which is not a Banach space.

The central limit theorem for random sets is

THEOREM: Let $X_1, X_2, \ldots, X_k, \ldots$ be i.i.d. random sets with $E\|X_1\|^2 < \infty$. Then the limiting distribution of

$$\sqrt{k}\, d\left(\frac{1}{k}(X_1 + X_2 + \cdots + X_k), E(\text{co } X_1)\right)$$

as $k \to \infty$ is the distribution of $\|Z\|$, where Z is a Gaussian variable valued in $C(S^{n-1})$. In this statement d is the Hausdorff distance.

The concept of limiting distribution is needed here in a particular context. Suppose $\eta_1, \eta_2, \ldots$ are positive random variables. To say that the limiting distribution of η_k as $k \to \infty$ is η means that

$$P\{\eta_k \leq x\} \to P\{\eta \leq x\}$$

for every real number x.

In essence, the message of this theorem is that one can approximately calculate probabilities related to the distance between the average $k^{-1}(X_1 + X_2 + \cdots + X_k)$ and the expected value $E(\text{co } X_1)$.

STRONG LAW OF LARGE NUMBERS FOR FUZZY RANDOM VARIABLES

A fuzzy random variable is a measurable function $X: \Omega \to F(R^n)$. If $F(R^n)$ is not separable, it is possible that the sum of two fuzzy random variables is no longer a fuzzy random variable, which means that it is no longer measurable.

What is this separability after all? Recall the situation of real numbers R and rational numbers Q. Clearly, $Q \subset R$, and it is well known that every real number can be approximated as close as we wish by rational numbers. For example, $\sqrt{2}$ is the limit of rationals 1.4, 1.41, 1.414, 1.4142, ... Note that the set Q is "rather small" as compared with R. Indeed, Q is countable (i.e., all its elements can be written in a sequence $r_1, r_2, \ldots, r_n$, ...). Therefore we say that R is *separable*. Similarly, R^n is separable because every vector can be approximated as closely as we want by a vector in the countable subset Q^n.

It turns out that the space $K(R^n)$ of compact sets is separable with respect to the Hausdorff distance There exists a countable family of sets in $K(R^n)$, $C \subset K(R^n)$, such that, for every $W \in K(R^n)$ and every $\epsilon > 0$, there exists $c \in C$ with $d(W, c) < \epsilon$.

It is not necessary at this point to specify the family C. Since our fuzzy random variables take values in $F(R^n)$, we are now interested in the separability of the space $F(R^n)$. Recall that we have introduced several distances in $F(R^n)$. The most important for our discussion are d_∞ and d_1. It is precisely because of separability properties that we prefer one distance over another.

It is easy to show that the space $(F(R^n), d_\infty)$ is not separable. The space $(F(R^n), d_1)$, on the other hand, is separable (although this is not so easy to show). Therefore, for the purpose of establishing a law of large numbers, the metric d_1 is preferable.

If we insist on using d_∞, then we have to restrict the space $F(R^n)$ to a subspace of it by imposing additional geometric restrictions on the membership functions we consider. This will be our approach in the next section, devoted to the central limit theorem.

The fuzzy random variables $X_1, X_2, \ldots, X_k$ are called *independent* if

$$P\{X_1 \in U_1, X_2 \in U_2, \ldots, X_k \in U_k\}$$
$$= P\{X_1 \in U_1\} P\{X_2 \in U_2\} \cdots P\{X_k \in U_k\}$$

for all open sets $U_1, U_2, \ldots, U_k \subseteq F(R^n)$. By open sets we mean a concept similar to that given in the previous section for $K(R^n)$.

The fuzzy random variables in a sequence $X_1, X_2, \ldots, X_k, \ldots$ are said to be *independent* if the preceding property holds for every finite collection of fuzzy random variables in this sequence. $X_1, X_2, \ldots, X_k, \ldots$ are *identically distributed* if

$$P\{X_1 \in U\} = P\{X_2 \in U\} = \cdots = P\{X_k \in U\} = \cdots$$

for every open $U \subseteq F(R^n)$.

Recall that the *support* of a fuzzy random variable is the random set supp X defined by

$$(\text{supp } X)(\omega) = \text{supp}[X(\omega)] \qquad \omega \in \Omega$$

The last concept we have to clarify before stating our main result is that of a convex hull of a fuzzy set. A fuzzy set $v: R^n \to [0, 1]$ is called a *fuzzy convex set* if

$$v(\lambda x + (1 - \lambda)y) \geq \min[v(x), v(y)]$$

for every $x, y \in R^n$ and every $\lambda \in [0, 1]$.

Clearly, if $C \subset R^n$ is a convex set, then χ_C is a fuzzy convex set; so fuzzy convexity of sets generalizes ordinary convexity of sets. It is easy to prove that v is a fuzzy convex set if and only if $L_\alpha v$ is a convex set in R^n for every $0 < \alpha < 1$.

The idea behind fuzzy convexity is this: If $v(x) = v(y) = 1$, then the defining inequality will force $v(\lambda x + (1 - \lambda)y) = 1$. If $u \in F(R^n)$ is not necessarily fuzzy convex, then its *convex hull* co u is the smallest fuzzy convex set containing u.

$$\text{co } u = \inf\{v \in F(R^n)/v \text{ is fuzzy convex}, v \supseteq u\}$$

It is possible to show that our approximation along the levels is consistent with this definition:

$$L_\alpha(\text{co } u) = \text{co}(L_\alpha u) \qquad 0 \leq \alpha \leq 1$$

To every fuzzy random variable $X: \Omega \to F(R^n)$, we can associate its convex hull co X defined by

$$(\text{co } X)(\omega) = \text{co}[X(\omega)] \qquad \omega \in \Omega$$

We are now ready to state the strong law of large numbers for fuzzy random variables.

THEOREM: Let $X_1, X_2, \ldots, X_k, \ldots$ be i.i.d. fuzzy random variables such that $E\|\text{supp } X_1\| < \infty$. Then

$$\lim_{k \to \infty} d_1\left(\frac{X_1 + X_2 + \cdots + X_k}{k}, E(\text{co } X_1)\right) = 0$$

almost surely.

This theorem can be proved if we assume that the fuzzy random variables take fuzzy convex values. The subspace of $F(R^n)$ consisting of fuzzy convex sets can be embedded into a Banach space. Then we apply a law of large numbers for Banach space valued random variables. Finally, the fuzzy convexity is dropped by a device known as the Shapley-Folkman lemma, which is extended for this context. Details can be found in (Klement, Puri, and Ralescu, 1986).

Intuitively, this theorem shows that the average of fuzzy random variables converges to the constant fuzzy set $E(\operatorname{co} X_1)$ (average is convexifying!), the convergence being in the metric d_1 (which is a sort of average of convergence along the level sets).

CENTRAL LIMIT THEOREM FOR FUZZY RANDOM VARIABLES

In this section, mainly because of some technical difficulties, we have to restrict the space of values of our fuzzy random variables. Instead of $F(R^n)$, we will consider its subspace $F_0(R^n)$ consisting of fuzzy sets $u \in F(R^n)$ with the following property: There exists a constant $M > 0$ such that

$$d(L_\alpha u, L_\beta u) \le M|\alpha - \beta|$$

for every $\alpha, \beta \in [0, 1]$.

We also say that the map $\alpha \to L_\alpha u$ satisfies a Lipschitz condition (with respect to the Hausdorff metric).

Example 1

If $u \in F(R^n)$ and if there is a constant $M > 0$ such that

$$|u(x) - u(y)| \ge M\|x - y\|$$

for all $x, y \in \operatorname{supp} u$, then $u \in F_0(R^n)$.

The above restriction implies that every membership function must be a one-to-one function.

Example 2

Let $u \in F(R^n)$. If

$$\min_{1 \le j \le n} \inf\left\{\left|\frac{\partial u}{\partial x_j}(x)\right| \, x \in \operatorname{supp} u \setminus L_1 u\right\} > 0$$

then $u \in F_0(R^n)$. Here the infimum is taken over all points of $\operatorname{supp} u \setminus L_1 u$ where the derivatives of u exist.

It is easier to understand this condition in the one-dimensional case, $F(R)$.

Here, if the derivative of u (where it exists) is bounded away from zero on supp $u \setminus L_1 u$, then u will belong to the space $F_0(R)$.

Another important space in connection with the central limit theorem is $C([0, 1] \times S^{n-1})$. Recall that S^{n-1} is the unit sphere in R^n. The product $[0, 1] \times S^{n-1}$ is a cylinder in R^{n+1} (indeed, if $n = 2$, then S^1 is the unit circle in the plane and $[0, 1] \times S^1$ is a cylinder in the space).

By $C([0, 1] \times S^{n-1})$ we mean all continuous functions

$$\varphi : [0, 1] \times S^{n-1} \to R$$

The norm of such a function is defined by

$$\|\varphi\| = \sup_{\substack{0 \leq \alpha \leq 1 \\ \|x\| = 1}} |\varphi(\alpha, x)|$$

It is not hard to show that $C([0, 1] \times S^{n-1})$ is a Banach space.

The concept of a gaussian element with values in $C([0, 1] \times S^{n-1})$ is similar to the corresponding concept where the space of values was $C(S^{n-1})$, described in the previous chapter. So

$$Z : \Omega \to C([0, 1] \times S^{n-1})$$

is *Gaussian* if for every $(\alpha_1, t_1), \ldots, (\alpha_r, t_r) \in [0, 1] \times S^{n-1}$ the distribution of the random vector $(Z(\alpha_1, t_1), Z(\alpha_2, t_2), \ldots, Z(\alpha_r, t_r))$ is multivariate normal (in R^r).

The central limit theorem for fuzzy random variables is

THEOREM: Let $X_1, X_2, \ldots : \Omega \to F_0(R^n)$ be i.i.d. fuzzy random variables satisfying

(a) $E \|\operatorname{supp} X_1\|^2 < \infty$.

(b) $E \left[\sup_{\alpha \neq \beta} \dfrac{d(L_\alpha X_1, L_\beta X_1)}{|\alpha - \beta|} \right]^2 < \infty$.

Then there exists a Gaussian random element Z in $C([0, 1] \times S^{n-1})$ such that the limiting distribution of

$$\sqrt{k} \, d_\infty \left(\frac{X_1 + X_2 + \cdots + X_k}{k}, E(\operatorname{co} X_1) \right)$$

is the distribution of $\|Z\|_\infty$.

This result is more complicated than the central limit theorem for random sets. Here we have the finiteness condition for the second moment (a), but we also have the additional hypothesis (b). It is also a finiteness of a second moment but with respect to a different norm.

The conclusion is similar to that for random sets. Note that here we have convergence in d_∞, not d_1. The reason is that we are restricted to the subspace $F_0(R^n)$ of $F(R^n)$. It turns out that $(F_0(R^n), d_\infty)$ is separable, whereas $(F(R^n), d_\infty)$ is not.

The proof is similar to that for random sets. First we restrict our attention to fuzzy convex-valued fuzzy random variables. Then we apply a central limit theorem for Banach space valued random variables, since the space $F_0(R^n)$ can be *embedded* into a Banach space. Finally, the fuzzy convexity is dropped by using an extension of the Shapley-Folkman lemma. Details can be found in the bibliography.

Chapter 6

FUZZY SET-VALUED MEASURES

The importance of the concept of measure in statistics is well known. The concept of a probability measure is fundamental to the development of the field. That probabilities are known numbers is essential for statistical analysis.

However, probabilities are obtained by a particular case of measurement, and the measurement process is imprecise, so the values are known only with some tolerance. This suggests the possibility of taking the values of the measure (in particular, probability measure) to be sets (intervals, in the real-valued case) or, more generally, fuzzy sets. This chapter suggests that this point of view is a possible approach to the problem of inexact measurement.

We will first introduce the concept of a set-valued measure, the values being compact convex subsets of R^n. We will also define the expected value of a random variable with respect to a set-valued measure. Then we will introduce and study set-valued measures and expected values with respect to such measures.

Note that the problems to be approached here are, in some sense, dual to those studied in the previous chapters. Indeed, there the variables were set-valued, and the measures were point-valued. Here we consider set-valued measures and point-valued variables. Such a theory is very useful in the next chapters, devoted to statistical analysis with imprecise information.

SET-VALUED MEASURES

Recall that $K(R^n)$ denotes the space of all nonempty compact subsets of R^n, and $K_c(R^n)$ denotes the space of all compact convex nonempty subsets of R^n. Since $K(R^n)$ is a metric space (with the Hausdorff metric), it makes sense to talk about series $\Sigma_{j=1}^{\infty} W_j$ of sets W_j in $K(R^n)$. More precisely, a series $\Sigma_{j=1}^{\infty} W_j$ is said to be *convergent* if there exists a set $W \in K(R^n)$ such that

$$\lim_{P \to \infty} d\left(\sum_{j=1}^{P} W_j, W \right) = 0$$

A series $\Sigma_{j=1}^{\infty} W_j$ is said to be *unconditionally convergent* if the series of positive numbers $\Sigma_{j=1}^{\infty} \|W_j\|$ converges. Recall that the norm of a set $W \in K(R^n)$ is

$$\|W\| = d(W, \{0\}) = \sup_{a \in W} \|a\|$$

As usual, Ω denotes a set, the totality of outcomes of a random experiment, and $\mathcal{A}$ denotes a σ-algebra of subsets of Ω (the collection of measurable events).

Definition

A *set-valued measure* is a set function

$$\mu: \mathcal{A} \to K(R^n)$$

with the properties

(a) $\mu(\emptyset) = 0$,
(b) $\mu(\bigcup_{j=1}^{\infty} A_j) = \Sigma_{j=1}^{\infty} \mu(A_j)$,
for every sequence $(A_j)_j$, $A_j \in \mathcal{A}$ such that $A_i \cap A_j = \emptyset$ for $i \neq j$.

Note that these conditions are completely similar to the corresponding conditions used in the definition of a measure. The right side of (b) is a series of compact subsets of R^n, and it is *implicitly* assumed that this series is unconditionally convergent.

Roughly speaking, a set-valued measure is a function that assigns a set as the measure of an event in $\mathcal{A}$. Property (b) is also referred to as countable additivity.

In the following, for simplicity, we will assume that our set-valued measures take on values in $K_c(R^n)$; that is, the values are compact, convex, and nonempty.

Our main goal now is to define the expected value of a random variable with respect to a set-valued measure (also called the integral with respect to a set-valued measure). This can be done in three equivalent ways. For a better understanding of this concept, we sketch these methods. As usual, we do not give formal proofs since they can be found in the bibliography.

Method 1

Let $X: \Omega \to R$ be a random variable (i.e., measurable), and let $\mu: \mathcal{C} \to K_c(R^n)$ be a set-valued measure. We want to define $\int_\Omega X \, d\mu$ as a compact convex nonempty set. [i.e., an element of $K_c(R^n)$]. Suppose first that $X = \sum_{j=1}^k \alpha_j \chi_{A_j}$ is a simple function. Define

$$\int_\Omega X \, d\mu = \sum_{j=1}^k \alpha_j \mu(A_j)$$

and, obviously, $\int_\Omega X \, d\mu$ is a set.

If X is measurable, there exists a sequence φ_n of simple functions such that $\varphi_n \to X$ (actually more conditions are needed here, but they are not essential at this point). We set

$$\int_\Omega X \, d\mu = \lim_{n \to \infty} \int_\Omega \varphi_n \, d\mu$$

where the limit is in the Hausdorff distance.

Of course, there are points that need further clarification (how do we know that the sets $\int_\Omega \varphi_n \, d\mu$ converge?), but basically this is the main idea behind the definition of the integral by this method.

Method 2

This method defines the integral with respect to a set-valued measure by using the concept of a selector. A selector ν of a set-valued measure μ is a vector-valued measure

$$\nu: \mathcal{C} \to R^n$$

such that $\nu(A) \in \mu(A)$ for every $A \in \mathcal{C}$.

Selectors do exist, although this is by no means a trivial problem. If $X: \Omega \to R$, we define $\int_\Omega X \, d\mu$ as

$$\int_\Omega X \, d\mu = \left\{ \int_\Omega X \, d\nu \,\middle|\, \nu \text{ is a selector of } \mu \right\}$$

This definition has the advantage of being simple, but, it is harder to study properties of the integral in this way.

Method 3

This is a more difficult approach to the concept of integral, but once it is developed it becomes easier to study its properties. The first step of this approach is to realize that the space $K_c(R^n)$ (which is not a Banach space, since it is not even a vector space) can be embedded into a Banach space. The crucial fact is that this embedding is isometric; that is, it preserves the distance.

Recall that S^{n-1} denotes the unit sphere $\{x \in R^n \mid \|x\| = 1\}$ in R^n and that $C(S^{n-1})$ is the (Banach) space of all continuous functions from S^{n-1} into R. It is possible to associate to every compact convex set $W \in K_c(R^n)$ its *support function* $s_W: S^{n-1} \to R$, defined by

$$s_W(x) = \sup_{a \in W} \langle x, a \rangle$$

Here $\langle\ ,\ \rangle$ denotes the inner product in R^n (i.e., $\langle x, y \rangle = \sum_{i=1}^{n} x_i y_i$).

The support function induces the desired embedding of $K_c(R^n)$ into $C(S^{n-1})$. It is possible to show that there is an embedding

$$j: K_c(R^n) \to C(S^{n-1})$$

such that

(i) $j(W + T) = j(W) + j(T)$.
(ii) $j(\alpha W) = \alpha j(W)$ for $\alpha \geq 0$.
(iii) $\|j(W) - j(T)\| = d(W, T)$.

The embedding is defined by

$$j(W) = s_W$$

and property (iii) shows that j is an isometry (i.e., the distances are preserved).

What is the role of such an embedding here? The main point is that any set-valued measure $\mu: \mathcal{C} \to K_c(R^n)$ can be extended to a vector-valued measure $\hat{\mu}: \mathcal{C} \to C(S^{n-1})$ simply by the composition

$$\mathcal{C} \xrightarrow{\mu} K_c(R^n) \xrightarrow{j} C(S^{n-1}) \qquad \hat{\mu} = j \circ \mu$$

More exactly,

$$\hat{\mu}(A) = j(\mu(A)) = s_{\mu(A)} \qquad A \in \mathcal{C}$$

So $\hat{\mu}$ associates to every event $A \in \mathcal{C}$ the support function of $\mu(A)$.

Assume now that $X: \Omega \to R$, $X \geq 0$, is a positive random variable (this method works best for such variables). First, consider the integral $\int_\Omega X \, d\hat{\mu}$ with respect to $\hat{\mu}$; this makes sense because $\hat{\mu}$ has values in a nice Banach space. Clearly, $\int_\Omega X \, d\hat{\mu} \in C(S^{n-1})$. We want to come up with a set in $K_c(R^n)$. Not every continuous function in $C(S^{n-1})$ comes from a set via the support function. In other words, there are many $\varphi \in C(S^{n-1})$ for which there is no $W \in K_c(R^n)$ such that $\varphi = s_W$. Is $\int_\Omega X \, d\hat{\mu}$ such a function?

It can be shown that where $\int_\Omega X \, d\hat{\mu}$ is concerned, there exists a unique set $W \in K_c(R^n)$ such that

$$\int_\Omega X \, d\hat{\mu} = s_W$$

By *definition*, this compact convex set W is taken to be the integral of X with respect to μ (i.e., $\int_\Omega X \, d\mu$). So the integral is defined by the property

$$\int_\Omega X \, d\hat{\mu} = s_{\int_\Omega X \, d\mu}$$

Roughly speaking, the support of the integral is the integral with respect to the support measure.

It is possible to show that the integrals defined by these three methods are equivalent.

Another important point is this: The integral $\int_\Omega X \, d\mu$ may not always exist. What is the integrability condition? To explain this, note that to each set-valued measure $\mu: \mathcal{C} \to K_c(R^n)$ we can associate its total variation $\|\mu\|$. This is a positive (real-valued) measure defined by

$$\|\mu\|(A) = \sup \sum_{j=1}^\infty \|\mu(A_j)\|$$

where the supremum is taken over all (disjoint) partitions $(A_j)_j$ of A.

The condition under which $\int_\Omega X \, d\mu$ exists is precisely $\int_\Omega X \, d\|\mu\| < \infty$ (or $\int_\Omega |X| \, d\|\mu\| < \infty$, if X is not positive). Note that $\int_\Omega X \, d\|\mu\|$ is a classical integral since both X and $\|\mu\|$ are real valued.

Now we are ready to study some properties of the integral we have just introduced. The simpler ones are collected in the following proposition.

PROPOSITION: Let $X, Y: \Omega \to R$, $X, Y \geq 0$ and let $c \geq 0$ be a constant. Then

(a) $\int_\Omega (X + Y) \, d\mu = \int_\Omega X \, d\mu + \int_\Omega Y \, d\mu$.
(b) $\int_\Omega (cX) \, d\mu = c \int_\Omega X \, d\mu$.
(c) $\|\int_\Omega X \, d\mu\| \leq \int_\Omega X \, d\|\mu\|$.

Another important property is a theorem of the Lebesgue dominated convergence type for the integral with respect to a set-valued measure. This theorem gives conditions under which the limit and the integral can be permuted.

THEOREM: Let X_k, $Y \geq 0$, and assume that $\lim_{k \to \infty} |X_k - X| = 0$, $X_k \leq Y$ for each $k = 1, 2, \ldots$, and $\int_\Omega Y \, d\|\mu\| < \infty$. Then

$$\int_\Omega X \, d\|\mu\| < \infty$$

$$\lim_{k \to \infty} d\left(\int_\Omega X_k \, d\mu, \int_\Omega X \, d\mu\right) = 0$$

The message of this result is that out of $X_k \to X$ we can infer $\int_\Omega X_k \, d\mu \to \int_\Omega X \, d\mu$ (in the Hausdorff metric) as long as X_k is dominated by a function Y, integrable with respect to $\|\mu\|$.

FUZZY SET-VALUED MEASURES

In this section we consider the space $F_c(R^n)$ of all fuzzy sets $u: R^n \to [0, 1]$ such that $L_\alpha u$ is compact and convex for each $0 < \alpha \leq 1$, supp u is compact, and $L_1 u \neq \emptyset$.

The distance to be used here is

$$d_\infty(u, v) = \sup_{\alpha > 0} d(L_\alpha u, L_\alpha v)$$

Recall that $(F_c(R^n), d_\infty)$ is a complete metric space. Series of fuzzy sets (i.e., infinite sums) can be approached as follows. We say that $\Sigma_{j=1}^\infty u_j$ is *unconditionally convergent* if $\Sigma_{j=1}^\infty \|u_j\|$ is finite.

A *fuzzy set-valued measure* will be a set function valued in $F_c(R^n)$ that has the property of countable additivity.

Definition

A *fuzzy set-valued measure* is a set function

$$\mu: \mathcal{C} \to F_c(R^n)$$

such that

(a) $\mu(\emptyset) = \chi_{\{0\}}$,
(b) $\mu(\bigcup_{j=1}^\infty A_j) = \Sigma_{j=1}^\infty \mu(A_j)$
   for every collection $(A_j)_j$ of mutually disjoint events.

As with set-valued measures, in the countable additivity condition (b) it is implicit that $\Sigma_{j=1}^\infty \|\mu(A_j)\| < \infty$; that is, that the series $\Sigma_{j=1}^\infty \mu(A_j)$ of *fuzzy sets* is unconditionally convergent.

Note: For simplicity, in this section we use notations similar to those of the previous section, although we are now talking about fuzzy set-valued measures as an extension of set-valued measures.

At this stage we would like to define the integral with respect to a fuzzy set-valued measure. In some sense, this concept is dual to the expectation of a fuzzy random variable. Our crucial tool will again be the representation theorem for fuzzy sets.

If $X: \Omega \to R$ is a real-valued random variable, we would like to define $\int_\Omega X \, d\mu$ so that

$$L_\alpha \left(\int_\Omega X \, d\mu \right) = \int_\Omega X \, d\mu_\alpha \quad \alpha \in (0, 1]$$

where μ_α is a level α set-valued measure that will be defined later. We have to emphasize that a level condition such as the previous one is motivated by the fact that levels approximate fuzzy sets.

Suppose $\mu: \mathcal{Q} \to F_c(R^n)$, and let $0 < \alpha \leq 1$. The α-level of μ is the set-valued measure $\mu_\alpha: \mathcal{Q} \to K_c(R^n)$ defined by

$$\mu_\alpha(A) = L_\alpha[\mu(A)]$$

Actually, to show that μ_α is indeed a set-valued measure (countable additivity is what we have in mind), we need the following property of series of fuzzy sets:

$$L_\alpha \left(\sum_{j=1}^\infty u_j \right) = \sum_{j=1}^\infty L_\alpha u_j \quad 0 < \alpha \leq 1$$

provided that $\sum_{j=1}^\infty u_j$ is an unconditionally convergent series of fuzzy sets in $F_c(R^n)$.

This property does indeed hold, and although we do not prove it here, it has to do with the completeness of the space $(F_c(R^n), d_\infty)$ as mentioned earlier.

The *support* of a fuzzy set-valued measure μ is the set-valued measure supp μ such that

$$(\text{supp } \mu)(A) = \text{supp}[\mu(A)] \quad A \in \mathcal{Q}$$

It is not trivial to show that supp μ is indeed a set-valued measure. In the right side, supp denotes the support of the corresponding fuzzy set (i.e., the closure of the set of points whose membership degrees are not zero).

We are now in position to define the integral with respect to μ. For simplicity, let us consider only positive random variables $X: \Omega \to R, X \geq 0$.

Definition

The *integral* of X with respect to the fuzzy set-valued measure μ, denoted by $\int_\Omega X \, d\mu$, is the unique fuzzy set satisfying

$$L_\alpha \left(\int_\Omega X \, d\mu \right) = \int_\Omega X \, d\mu_\alpha$$

for every $0 < \alpha \leq 1$.

How do we know that the integral exists? Actually it does not unless we require that

$$\int_\Omega X \, d \| \operatorname{supp} \mu \| < \infty$$

which is our integrability condition. Once this condition is imposed, by using *the representation theorem*, one can prove existence. We give a sketch of the proof: Fix $0 < \alpha \leq 1$ and define the set $W_\alpha = \int_\Omega X \, d\mu_\alpha$. Then show that the family of sets $(W_\alpha)_\alpha$ satisfies the hypotheses of the representation theorem (i.e., $\alpha \leq \beta \Rightarrow W_\alpha \supseteq W_\beta$ and the crucial "continuity" condition $\alpha_1 \leq \alpha_2 \leq \cdots$, $\lim_{k \to \infty} \alpha_k = \alpha \Rightarrow W_\alpha = \bigcap_{k=1}^\infty W_{\alpha_k}$). Then we find the fuzzy set u (provided by the representation theorem) such that $L_\alpha u = W_\alpha$, $0 < \alpha \leq 1$. This set (whose uniqueness is quite clear in view of the level requirements) is our integral $\int_\Omega X \, d\mu$ with respect to the fuzzy set-valued measure μ.

The message of this discussion is that behind the foregoing definition there is a complicated mathematical proof that ensures the existence of the object we consider (the integral in this case).

The practical computation of the integral $\int_\Omega X \, d\mu$ could be done in different ways. Perhaps the easiest way would be to start with simple functions $X = \sum_{j=1}^k \alpha_j \chi_{A_j}$ and the formula

$$\int_\Omega X \, d\mu = \sum_{j=1}^k \alpha_j \, \mu(A_j)$$

If X is more complicated, consider a sequence of simple functions φ_n such that $\varphi_n \to X$ and define

$$\int_\Omega X \, d\mu = \lim_{n \to \infty} \int_\Omega \varphi_n \, d\mu$$

Some theorems are needed here, first to show under what conditions this calculation is correct, and, second, to show that what we obtain is the integral given by the definition.

Some properties of this new integral are given next; they obviously generalize the corresponding properties of the integral with respect to a set-valued measure, discussed earlier.

PROPOSITION: Let $X, Y \geq 0$ be random variables and $c \geq 0$ a constant. Assume that $\int_\Omega X \, d \| \operatorname{supp} \mu \| < \infty$ and $\int_\Omega Y \, d \| \operatorname{supp} \mu \| < \infty$. Then

(a) $\int_\Omega (X + Y)\,d\mu = \int_\Omega X\,d\mu + \int_\Omega Y\,d\mu$.
(b) $\int_\Omega (cX)\,d\mu = c \int_\Omega X\,d\mu$.
(c) $\| \int_\Omega X\,d\mu \| \le \int_\Omega X\,d\| \operatorname{supp} \mu \|$.

As the reader might guess, we can also prove a theorem of the Lebesgue dominated convergence type. This result is stated next without proof.

THEOREM: Let X_k, $Y \ge 0$ and assume that $X_k \le Y$ for $k = 1, 2, \ldots$ with $\int_\Omega Y\,d\| \operatorname{supp} \mu \| < \infty$. If $\lim_{k \to \infty} |X_k - X| = 0$ (pointwise), then

$$\int_\Omega X\,d\| \operatorname{supp} \mu \| < \infty$$

$$\lim_{k \to \infty} d_\infty \left(\int_\Omega X_k\,d\mu, \int_\Omega X\,d\mu \right) = 0$$

STRONG LAW OF LARGE NUMBERS AND FUZZY SET-VALUED MEASURES

In this section we consider a sequence of i.i.d. random variables with respect to a fuzzy set-valued measure. In such a framework we state a law of large numbers.

Although we postpone giving more details on statistical inference with inexact data, we note here the main motivation behind it. The main idea of any Bayesian approach to statistical analysis is the presence of prior information. Some information is available about the unknown parameters before the statistical experiment is performed. The classical Bayesian approach assumes that this prior information is given by a probability density on the parameter space.

However, in many instances the prior information is imprecise or not totally reliable. Many alternative approaches have been proposed, including vague priors, upper and lower probabilities, fuzzy measures, and so on. If we consider that the prior information is in the form of a fuzzy set-valued measure, then the Bayes risk associated to any estimator (this is the average risk with respect to the prior) leads to an integral with respect to such a measure.

In this section we will consider a *fuzzy set-valued probability measure* as a set function $\pi: \mathcal{Q} \to F_c([0,1])$ with the properties:

(a) $\pi(\emptyset) = \chi_{\{0\}}$.
(b) $\pi(\bigcup_{j=1}^\infty A_j) = \sum_{j=1}^\infty \pi(A_j)$ if $(A_j)_j$ are mutually disjoint events in $\mathcal{Q}$.
(c) $\pi(\Omega)(1) = 1$.

Property (c) states that the fuzzy set $\pi(\Omega)$ "contains" 1.

In the rest of this section we consider a fuzzy set-valued probability space as a triple $(\Omega, \mathcal{A}, \pi)$, where Ω is a set, $\mathcal{A}$ is a σ-algebra of subsets of Ω, and π is a fuzzy set-valued probability measure on $\mathcal{A}$.

Let $X: \Omega \to R^n$ be a random vector (i.e., an n-tuple of random variables). This implicitly means that X is measurable, that is, $\{X \in U\} \in \mathcal{A}$ for every open set U in R^n.

Such a random vector induces a fuzzy set-valued probability measure in R^n, denoted by

$$\pi_X(B) = \pi\{X \in B\}$$

for subsets B of R^n (actually Borel subsets).

If we have a sequence of random vectors $X_1, X_2, \ldots$, we say that they are *identically distributed* if $\pi_{X_1} = \pi_{X_2} = \cdots$. This means that

$$\pi\{X_1 \in B\} = \pi\{X_2 \in B\} = \cdots$$

for every B.

The definition of independence requires more care. After all, the values of the "probability" π are fuzzy sets and not numbers. For this reason we have to define the product of fuzzy subsets of $[0, 1]$.

Suppose that $u, v: [0, 1] \to [0, 1]$. Their *product* is

$$(uv)(x) = \sup_{yz=x} \min[u(y), v(z)] \qquad 0 \le x \le 1$$

This is an extension of the product of sets $M, N \subseteq [0, 1]$:

$$MN = \{mn \mid m \in M, n \in N\}$$

We say that the random vectors $X_1, \ldots, X_k$ are *independent* (with respect to π) if

$$\pi\{X_1 \in B_1, \ldots, X_k \in B_k\}$$
$$= \pi\{X_1 \in B_1\} \pi\{X_2 \in B_2\} \cdots \pi\{X_k \in B_k\}$$

for all $B_1, B_2, \ldots, B_k$ (Borel) subsets of R^n. Note that in the right side we have the product of fuzzy sets.

The random vectors in the sequence $X_1, X_2, \ldots$ are *independent* if every finite subcollection of random vectors is independent.

We will also assume that π is such that supp π is "dominated" by a genuine probability measure P. We write this as

$$\text{supp } \pi \ll P$$

with the meaning that for every event $A \in \mathcal{A}$ for which $P(A) = 0$ we also have supp $\pi(A) = \{0\}$. This is a technical assumption that can be made without any loss of generality.

Some more notation is in order before we can state the law of large numbers. This concerns the distance from a point to a fuzzy set. First, if $B \subset R^n$ and $x \in R^n$, we set

$$\text{dist}(x, B) = \inf_{a \in B} \|x - a\|$$

so dist(x, B) is the shortest distance from x to B.

If $u: R^n \to [0, 1]$ is a fuzzy set, the *distance* from x to u, which seems to be best suited for our purposes here, is

$$d_1(x, u) = \int_0^1 \text{dist}(x, L_\alpha u) \, d\alpha$$

Note that this resembles the distance d_1 introduced earlier. However, d_1 here does not have anything to do with Hausdorff distance. Actually, as will be clear from the substance of our main result, the Hausdorff distance and its generalizations are not well suited for our present problem.

The law of large numbers concerns a sequence $X_1, X_2, \ldots$ of i.i.d. random vectors and the "convergence" of $k^{-1}(X_1 + X_2 + \cdots + X_k)$ to $\int_\Omega X_1 \, d\pi$. What does such a "convergence" mean? After all, $k^{-1}(X_1 + X_2 + \cdots + X_k)$ is a random vector in R^n, and $\int_\Omega X_1 \, d\pi$, the expected value with respect to π, is a fuzzy subset of R^n. The precise result, our strong law of large numbers with respect to a fuzzy set-valued probability measure, is

THEOREM: Let $X_1, X_2, \ldots$ be i.i.d. random vectors defined on a fuzzy set-valued probability space $(\Omega, \mathcal{C}, \pi)$ such that supp $\pi \ll P$, where P is a probability measure. Assume that $\int_\Omega \|X_1\| \, dP < \infty$. Then

$$d_1\left(\frac{1}{k}(X_1 + X_2 + \cdots + X_k), \int_\Omega X_1 \, d\pi\right) \to 0$$

almost surely with respect to π.

This theorem says that if k is large, the vector $k^{-1}(X_1 + X_2 + \cdots + X_k)$ has the tendency to "enter" the fuzzy set $\int_\Omega X_1 \, d\pi$.

Example

The law of large numbers was given for a fuzzy set-valued probability with values in $F_c(R^n)$. Actually the result is true if the values of π are just fuzzy subsets of $[0, 1]$. Under some mild additional hypotheses about π, it follows that its values must be fuzzy convex.

To gain more insight into fuzzy set-valued probabilities, let us note first that, since $\pi(A)$ is fuzzy convex, the levels $L_\alpha \pi(A)$ must be convex subsets of $[0, 1]$, that is, intervals.

Suppose $0 < \alpha \leq 1$ is fixed. Then

$$L_\alpha \pi(A) = \pi_\alpha(A)$$

and π_α is a set-valued measure. Since the values of π_α are in $K_c(R^n)$, it follows that these values must be closed intervals in $[0, 1]$.

It is possible to show that

$$\pi_\alpha(A) = [P_{1\alpha}(A), P_{2\alpha}(A)]$$

where $P_{1\alpha}$ is a measure, $P_{2\alpha}$ is a *probability* measure, and

$$P_{1\alpha}(A) \leq P_{2\alpha}(A) \quad \text{for every } A \in \mathcal{A}$$

So π_α is an interval of measures with the right endpoint a probability measure. Note that it would not be possible for both $P_{1\alpha}$ and $P_{2\alpha}$ to be probability measures. Actually, if P and Q are probability measures (on the same space) such that $P \leq Q$ (in the sense $P(A) \leq Q(A)$ for each $A \in \mathcal{A}$), then $P = Q$; that is, P and Q must coincide.

This is easy to show: Since $P(A) \leq Q(A)$, it follows that $P(\bar{A}) \leq Q(\bar{A})$, where $\bar{A}$ is the complement of A. But this gives $1 - P(A) \leq 1 - Q(A)$, so $P(A) \geq Q(A)$. This implies $P(A) = Q(A)$ for each A.

Now, if α, β are in $[0, 1]$ and $\alpha \leq \beta$, it follows that $\pi_\alpha \supseteq \pi_\beta$, so $[P_{1\alpha}(A), P_{2\alpha}(A)] \supseteq [P_{1\beta}(A), P_{2\beta}(A)]$. In particular, $P_{2\beta}(A) \leq P_{2\alpha}(A)$, but as we have already explained, since $P_{2\alpha}$ and $P_{2\beta}$ are both probability measures, it must be that $P_{2\alpha} = P_{2\beta}$. Thus the right endpoint of $\pi_\alpha(A)$ does not depend on α; that is, $\pi_\alpha(A) = [P_{1\alpha}(A), P_2(A)]$.

All this discussion implies something about the *structure* of $\pi(A)$, the values of our fuzzy set-valued probability. What can be concluded is that $\pi(A)$ is a fuzzy set on $[0, 1]$, and there exists an ordinary probability measure P_2 such that

$\pi(A)$ is nondecreasing on $[0, P_2(A)]$.
$\pi(A) = 0$ on $(P_2(A), 1]$.

Intuitively speaking, $\pi(A)$ is a fuzzy interval in $[0, 1]$ as in the example "much larger than 0.2 but smaller than 0.7."

Chapter 7

RELATIONSHIPS BETWEEN FUZZY RANDOM VARIABLES AND FUZZY SET-VALUED MEASURES

Let us briefly summarize what we have done in the previous three chapters. Classical probability theory is concerned with a framework dealing with random variables $X: \Omega \to R$ (or random vectors $X: \Omega \to R^n$) and a probability measure $P: \mathcal{C} \to [0, 1]$.

Both values of X and P are numbers (or X is a vector in R^n). The expected value EX is a sort of weighted average of the values of X, the weights being the corresponding probabilities. The quantity EX is a number (or a vector in R^n).

We first extended this framework by considering fuzzy random variables $X: \Omega \to F(R)$ [or $X: \Omega \to F(R^n)$]. The probability measure is still a function $P: \mathcal{C} \to [0, 1]$. So the values of X are fuzzy sets, and the values of P are numbers. The expected value is a fuzzy set, and a theory of fuzzy random variables was possible.

In Chapter 6 we extended the original framework of probability theory by taking a somewhat dual point of view. This time the random variables (or vectors) X assumed as values numbers (or vectors in R^n), and the values assumed by the measure were fuzzy sets. The expected value EX is again a fuzzy set.

These are two different approaches to the problem of combining probability and fuzziness. The reader has surely realized that there is some kind of duality between these approaches. It is exactly on this duality that we want to focus our attention in this chapter. The framework will be provided by the Radon-Nikodym theorem.

RADON-NIKODYM THEOREM FOR FUZZY SET-VALUED MEASURES

Assume that $(\Omega, \mathcal{C}, P)$ is a probability space, and let $X: \Omega \to F_c(R^n)$ be a fuzzy random variable. Can we come up with a fuzzy set-valued measure generated by X? The answer is yes, and this measure $\mu_X: \mathcal{C} \to F_c(R^n)$ is

$$\mu_X(A) = \int_A X\, dP \quad A \in \mathcal{C}$$

Note that we consider here the integral over A. Before, we used only the integral over the whole space Ω. This was our expected value EX. However, the integral over A can be defined in a completely similar way.

To be correct, μ_X makes sense only if X is "integrable." We have the following result.

PROPOSITION: If $E \| \text{supp } X \| < \infty$, then the set function

$$\mu_X(A) = \int_A X\, dP \quad A \in \mathcal{C}$$

is a fuzzy set-valued measure on $\mathcal{C}$ with the property

$$P(A) = 0 \quad \text{implies} \quad \mu_X(A) = \chi_{\{0\}}$$

Note that μ_X is "dominated" by P. We say that μ_X is *absolutely continuous* with respect to P (written $\mu_X \ll P$) since $P(A) = 0$ implies $\mu_X(A) = \chi_{\{0\}}$.

Now suppose that we want to go the other way around, that is, to start with a fuzzy set-valued measure $\mu: \mathcal{C} \to F_c(R^n)$ which is *absolutely continuous* with respect to P, $\mu \ll P$. The main question is, does there exist a fuzzy random variable X such that $\mu(A) = \int_A X\, dP$, $A \in \mathcal{C}$? In other words, can any measure dominated by P be recaptured in this way? As we shall see, the answer is yes. This is the Radon-Nikodym theorem for fuzzy set-valued measures.

We again mention that an affirmative answer to this question shows that each fuzzy set-valued measure "dominated" by P is representable as the integral of a suitable fuzzy random variable. This property will be used to show the meaning of the duality between the two approaches. More precisely, we show that the expected value of a random vector with respect to a fuzzy set-valued probability is equivalent to the expected value of a suitable fuzzy random variable.

Our main result, the theorem of Radon-Nikodym type, is

THEOREM: Let μ be a fuzzy set-valued measure on $\mathcal{A}$ such that $\mu \ll P$. There exists a fuzzy random variable $X: \Omega \to F_c(R^n)$ with $E \, \|\text{supp } X\| < \infty$ such that

$$\mu(A) = \int_A X \, dP \quad A \in \mathcal{A}$$

The proof (see bibliography) uses in a crucial way the representation theorem for fuzzy sets.

CONDITIONAL EXPECTATION OF A FUZZY RANDOM VARIABLE

Let $(\Omega, \mathcal{A}, P)$ be a probability space. Recall that Ω is the set of all outcomes of some random experiment, and $\mathcal{A}$ is a σ-algebra of subsets of Ω. Thus $\mathcal{A}$ is a collection of sets, the measurable events pertinent to the random experiment.

The Radon-Nikodym theorem can be used to introduce the important concept of conditional expectation of a fuzzy random variable. A fuzzy random variable is a function $X: \Omega \to F(R^n)$ measurable in the sense $\{\omega \in \Omega \mid X(\omega) \in U\} \in \mathcal{A}$ for every open subset $U \subseteq F(R^n)$ (in the structure of $F(R^n)$ generated by the Hausdorff metric).

At this point we consider a sub-σ-algebra $\mathcal{B} \subseteq \mathcal{A}$.

$\mathcal{B}$ is a subset of $\mathcal{A}$ (i.e., contains some events in $\mathcal{A}$) and a σ-algebra (i.e., closed with respect to complement and countable unions, containing the total space Ω).

We want to define the conditional expectation of a fuzzy random variable X with respect to the subalgebra $\mathcal{B}$. This will be denoted by $E(X|\mathcal{B})$.

In general, X will not be measurable with respect to $\mathcal{B}$; that is, it is not true that $\{X \in U\}$ belongs to $\mathcal{B}$ for every open U (as we have noticed, $\{X \in U\}$ always belongs to the larger σ-algebra $\mathcal{A}$). Note that $E(X|\mathcal{B})$ will have to be a fuzzy set-valued quantity.

Actually, what we are looking for is a new fuzzy random variable that will be measurable with respect to the smaller $\mathcal{B}$ and whose expected value is the same as that of X itself. Indeed, we want something more than that. We want the integral of the new fuzzy random variable over every set B to be the same as the integral of X over $B \in \mathcal{B}$. Expected values are a particular case of this situation because we take $B = \Omega \in \mathcal{B}$.

How can we do something like this? We start with X and $\mathcal{B} \subseteq \mathcal{A}$ and look for a new fuzzy random variable, $E(X|\mathcal{B})$, such that

$$\int_B E(X|\mathcal{B})\,dP = \int_B X\,dP$$

for every $B \in \mathcal{B}$.

Before describing the construction, it is important that we specify that the conditional expectation we are looking for, $E(X|\mathcal{B})$, is not a fuzzy set but a fuzzy random variable. At this stage we do not know if such an $E(X|\mathcal{B})$ exists. This will follow from the construction described next.

Define

$$\mu(B) = \int_B X\,dP \quad B \in \mathcal{B}$$

Obviously, this is a fuzzy set-valued measure defined on the σ-algebra $\mathcal{B}$. It is also obvious that $\mu \ll P$. We have to assume $E\,\|\operatorname{supp} X\| < \infty$; otherwise the integrals involved will not make sense.

These conditions being satisfied, we can apply the Radon-Nikodym theorem for fuzzy set-valued measures. So there exists a fuzzy random variable that we denote by $E(X|\mathcal{B})$ with the fundamental property that

$$\mu(B) = \int_B E(X|\mathcal{B})\,dP \quad B \in \mathcal{B}$$

This new fuzzy random variable we have obtained, $E(X|\mathcal{B})$, must be measurable with respect to the smaller σ-algebra $\mathcal{B}$.

In view of the original definition of the fuzzy set-valued measure and in consequence of the representation of μ by the Radon-Nikodym theorem, it follows that

$$\int_B E(X|\mathcal{B})\,dP = \int_B X\,dP \quad B \in \mathcal{B}$$

This is the concept we were looking for and we have shown that it exists.

Definition

Let X be a fuzzy random variable with $E\,\|\operatorname{supp} X\| < \infty$, and let $\mathcal{B} \subseteq \mathcal{A}$ be a sub-σ-algebra of $\mathcal{A}$. The *conditional expectation* of X with respect to $\mathcal{B}$, $E(X|\mathcal{B})$, is the fuzzy random variable satisfying the following properties:

(a) $E(X|\mathcal{B})$ is $\mathcal{B}$-measurable.
(b) $\int_B E(X|\mathcal{B})\,dP = \int_B X\,dP$ for every $B \in \mathcal{B}$.

Note, in particular, that property (b) implies $E(E(X|\mathcal{B})) = EX$ if we take $B = \Omega \in \mathcal{B}$.

As a particular case of conditional expectation, consider two fuzzy random variables X and Y. The σ-algebra generated by Y consists of all subsets of Ω of the form $Y^{-1}(U) = \{\omega \in \Omega \mid Y(\omega) \in U\}$, where U is an open set in $F_c(R^n_r)$. Clearly this σ-algebra, which we can denote by $\mathcal{B}$, is a sub-σ-algebra of $\mathcal{A}$ since Y is measurable.

Thus the conditional expectation $E(X \mid \mathcal{B})$ makes sense. Since $\mathcal{B}$ is generated by Y, we denote this conditional expectation by $E(X \mid Y)$, which is a fuzzy random variable.

FUZZY MARTINGALES

Statistics is based almost exclusively on a random sample $X_1, X_2, \ldots, X_k$, that is, on a finite collection of i.i.d. variables. The concept of *independence* is crucial for most of the theory. However, the assumption of independence is not universally valid, and there are situations where *dependence* rather than independence is the rule.

There are many different concepts of dependence for a sequence of random variables: for example, Markov chains, martingales, mixing sequences, and various special cases of them.

This section focuses on the particular concept of a martingale. Before going into definitions, we point out again that a fuzzy martingale will be a sequence of fuzzy random variables with a special kind of dependency relationship.

We start with a sequence of fuzzy random variables

$$X_1, X_2, \ldots, X_k, \ldots \qquad X_k: \Omega \to F_c(R^n)$$

and a sequence of sub-σ-algebras of $\mathcal{A}$, $\mathcal{B}_1, \mathcal{B}_2, \ldots, \mathcal{B}_k, \ldots$

Definition

The sequence of fuzzy random variables of σ-algebras $(X_k, \mathcal{B}_k)_k$ is called a fuzzy martingale if we have, for each $k = 1, 2, \ldots,$

(a) $\mathcal{B}_k \subseteq \mathcal{B}_{k+1}$.
(b) X_k is $\mathcal{B}_k$-measurable and $E \| \operatorname{supp} X_k \| < \infty$.
(c) $E(X_{k+1} \mid \mathcal{B}_k) = X_k$.

What this definition says is that the σ-algebras form an increasing sequence $\mathcal{B}_1 \subseteq \mathcal{B}_2 \subseteq \cdots \subseteq \mathcal{B}_k \subseteq \cdots$. Each X_k is measurable with respect to $\mathcal{B}_k$ (so also with respect to $\mathcal{B}_{k+1}, \mathcal{B}_{k+2}, \ldots$) and integrable. The crucial property is (c), which says that the conditional expectation of X_{k+1} with respect to $\mathcal{B}_k$ must be exactly X_k. Note that X_{k+1} is not $\mathcal{B}_k$-measurable but $\mathcal{B}_{k+1}$-measurable.

To get more insight into property (c), it helps if we think of $k = 1, 2, 3, \ldots$ as a moment of time. If we consider the fuzzy random variable X_{k+1} and we condition it by the information up to time k, given by $\mathcal{B}_k$, we must obtain the quantity X_k. It is therefore clear that a sequence $Y_1, Y_2, \ldots$ of i.i.d. fuzzy random variables is a fuzzy martingale.

We are interested in fuzzy martingales because we can have a convergence theorem generalizing, in a sense, the strong law of large numbers for fuzzy random variables.

THEOREM: Let $(X_k, \mathcal{B}_k)$ be a fuzzy martingale with $X_k: \Omega \to F_c(R^n)$. Assume that

$$\sup_{k \geq 1} E \, \|\operatorname{supp} X_k\| < \infty$$

Then there exists a fuzzy random variable $X: \Omega \to F_c(R^n)$ with $E \, \|\operatorname{supp} X\| < \infty$ such that $\lim_{k \to \infty} d_\infty(X_k, X) = 0$ almost surely.

The substance of this theorem is that every fuzzy martingale must converge almost surely in the metric d_∞ to an integrable fuzzy random variable.

Chapter 8

THE BAYES FORMULA FOR FUZZY PROBABILITIES

THE BAYES THEOREM FOR SET-VALUED MEASURES

A major chapter in statistical inference is based on the Bayesian point of view. The main idea behind this point of view is that of using prior information in an efficient way to make predictions about unknown parameters. Let us start with a set-valued probability measure, that is, a set-function

$$\pi: \mathcal{C} \to K([0, 1])$$

satisfying the properties

(i) $1 \in \pi(\Omega)$.
(ii) $\pi(\bigcup_{j=1}^{\infty} A_j) = \sum_{j=1}^{\infty} \pi(A_j)$ whenever $(A_j)_j$ is a disjoint collection of events in $\mathcal{C}$.

Recall that $\mathcal{C}$ is a σ-algebra of subsets of the space Ω, and $K([0, 1])$ stands for all nonempty closed subsets of $[0, 1]$. Let $M \in \mathcal{C}$ be a fixed event and $\mathcal{C}_M$ the restriction of $\mathcal{C}$ to M; that is, $\mathcal{C}_M = \{B \mid B = A \cap M$ for some $A \in \mathcal{C}\}$.

Intuitively, the idea is that we know for sure that the event M has occurred, and, in light of this information, we would like to reevaluate our probabilities of various events in $\mathcal{A}$. In other words, we would like to define a new set-valued probability on the *restricted* σ-algebra $\mathcal{A}_M$:

$$Q: \mathcal{A}_M \to K([0, 1])$$

such that

(a) $1 \in Q(M)$.
(b) $Q(\bigcup_{j=1}^{\infty} B_j) = \sum_{j=1}^{\infty} Q(B_j)$ whenever $(B_j)_j$ is a disjoint collection of events in $\mathcal{A}_M$.

It is clear (by analogy with point-valued probabilities) that Q should be proportional to π; that is,

$$Q(B) = \alpha \pi(B) \qquad B \in \mathcal{A}_M$$

for some proportionality *constant* α.

Since $1 \in Q(M) = \alpha \pi(M)$, it follows that $\alpha \sup \pi(M) = 1$, so $\alpha = 1/\sup \pi(M)$. Therefore

$$Q(B) = \alpha \pi(B) = \alpha \pi(A \cap M) = \frac{1}{\sup \pi(M)} \pi(A \cap M)$$

This new set-valued probability Q is the *conditional probability*, given $M \in \mathcal{A}$. We will denote it by $\pi(*|M)$. The preceding calculation gives the basic formula for the (set-valued) conditional probability

$$\pi(A|M) = \frac{1}{\sup \pi(M)} \pi(A \cap M)$$

If π is point-valued, one gets the usual formula for conditional probabilities.

Now suppose that we want to relate $\pi(M|A)$ to $\pi(A|M)$. The procedure is

$$\pi(M|A) = \frac{1}{\sup \pi(A)} \pi(M \cap A) = \frac{1}{\sup \pi(A)} \sup \pi(M) \pi(A|M)$$

$$= \frac{\sup \pi(M)}{\sup \pi(A)} \pi(A|M)$$

We are now ready to state the Bayes formula for set-valued probabilities.

THEOREM: Let $A_1, A_2, \ldots, A_n$ form a partition of the sample space Ω, and let $B \in \mathcal{A}$ be an event. Then

$$\pi(A_i|B) = \frac{\sup \pi(A_i)}{\sum_{j=1}^{n} \sup \pi(A_j) \sup \pi(B|A_j)} \pi(B|A_i) \qquad i = 1, 2, \ldots, n$$

Proof: First we get

$$\pi(A_i|B) = \frac{\sup \pi(A_i)}{\sup \pi(B)} \pi(B|A_i)$$

On the other hand,

$$\pi(B) = \pi\left[B \cap \left(\bigcup_{j=1}^{n} A_j\right)\right] = \sum_{j=1}^{n} \pi(B \cap A_j)$$

$$= \sum_{j=1}^{n} [\sup \pi(A_j)] \pi(B|A_j)$$

Note that if $U, V \subseteq [0, 1]$ and U, V are closed, then

$$\sup(U + V) = \sup U + \sup V$$

So

$$\sup \pi(B) = \sum_{j=1}^{n} \sup \pi(A_j) \sup \pi(B|A_j)$$

Finally we get

$$\pi(A_i|B) = \frac{\sup \pi(A_i)}{\sum_{j=1}^{n} \sup \pi(A_j) \sup \pi(B|A_j)} \pi(B|A_i)$$

which is the Bayes formula for set-valued probabilities.

Note that $\pi(A_i|B)$ is a scalar multiple of $\pi(B|A_i)$ and that the Bayes formula generalizes the corresponding classical formula where π is point valued.

THE CONCEPT OF INDEPENDENCE

How can one define independence of events with respect to a set-valued probability? If P is an ordinary probability measure, the events A and B are called independent if

$$P(A \cap B) = P(A)P(B)$$

One way of defining independence with respect to π is to extend this definition by using the product of subsets of $[0, 1]$:

$$UV = \{uv \,|\, u \in U, v \in V\} \quad U, V \subseteq [0, 1]$$

In the light of our definition of conditional set-valued probability, independence can be defined as follows. We call $A, B \in \mathcal{A}$ independent (with respect to π) if

$$\pi(A|B) = \pi(A) \quad \pi(B|A) = \pi(B)$$

This is equivalent to

$$\pi(A \cap B) = \sup \pi(A)\pi(B) = \sup \pi(B)\pi(A)$$

A particular case important in applications is determined by taking π to be compact convex valued or *interval valued*. Then it is easy to show that $\pi(A) = [Q(A), P(A)]$, where P is a probability measure and Q is a measure such that $Q \leq P$.

In this particular case, two events A and B are independent (with respect to π) if and only if the following conditions hold:

(a) $P(A \cap B) = P(A)P(B)$.
(b) $Q(A \cap B) = Q(A)P(B) = P(A)Q(B)$.

THE BAYES THEOREM FOR FUZZY SET-VALUED MEASURES

Situations of practical interest are those in which the values of the probabilities are not known exactly but can be expressed in linguistic terms, such as "very high," "approximately 0.9," "about $\frac{1}{2}$," or "near zero." Such situations can be modeled by using fuzzy set-valued probability measures. Such a measure is simply a function

$$\psi: \mathcal{A} \to F([0, 1])$$

satisfying

(a) $\psi(\Omega)(1) = 1$.
(b) $\psi(\bigcup_{j=1}^{\infty} A_j) = \sum_{j=1}^{\infty} \psi(A_j)$ whenever $(A_j)_j$ is a disjoint collection of events in $\mathcal{A}$.

Let $M \in \mathcal{A}$. We want to define the conditional probability $\psi(*|M)$. By analogy to what we did for set-valued probabilities, we set

$$\psi(A|M) = \alpha\psi(A \cap M)$$

where α is a constant. We can find α as follows: Let $A = M$. Then $\psi(M|M)$ is a fuzzy subset of $[0, 1]$ such that $\psi(M|M)(1) = 1$. So $[\alpha\psi(M)](1) = 1$. But this is equivalent to $\sup L_1(\alpha\psi(M)) = 1$, where $L_1 u = \{x | u(x) = 1\}$ is the 1-level set of the fuzzy set u.

Simple properties of level sets and of scalar multiplication give

$$\alpha \sup L_1\psi(M) = 1 \quad \text{so} \quad \alpha = \frac{1}{\sup L_1\psi(M)}$$

The conditional fuzzy probability formula becomes

$$\psi(A|M) = \frac{1}{\sup L_1\psi(M)} \psi(A \cap M)$$

In this context, the Bayes formula becomes

THEOREM (*Bayes formula for fuzzy probabilities*): Let $A_1, A_2, \ldots, A_n$ be a partition of the sample space Ω, and let $B \in \mathcal{C}$ be an event. Then

$$\psi(A_i|B) = \frac{\sup L_1\psi(A_i)}{\sum_{j=1}^{n} \sup L_1\psi(A_j) \sup L_1\psi(B|A_j)} \psi(B|A_i)$$

for $i = 1, 2, \ldots, n$.

The proof is similar to that for set-valued probabilities.

The concept of independence can also be extended to this framework. Let $\psi: \mathcal{C} \to F([0, 1])$ be a fuzzy probability measure. Two events $A, B \in \mathcal{C}$ are called independent with respect to ψ if

$$\psi(A|B) = \psi(A) \quad \psi(B|A) = \psi(B)$$

or, equivalently, if

$$\psi(A \cap B) = \sup [L_1\psi(A)]\psi(B) = \sup [L_1\psi(B)]\psi(A)$$

BIBLIOGRAPHY

A general framework for the manipulation of uncertainty in the design of knowledge-based systems can be found in

R. I. Goodman, and H. T. Nguyen, 1985, *Uncertainty Models for Knowledge Based Systems*, Amsterdam: North-Holland.

What we call fuzzy statistics is based on results to be found in

E. P. Klement, M. L. Puri, and D. A. Ralescu, 1984, Law of large numbers and central limit theorem for fuzzy random variables, in *Cybernetics and Systems Research* R. Trappl (ed.), Amsterdam: North-Holland.

E. P. Klement, M. L. Puri, and D. A. Ralescu, 1986, Limit theorems for fuzzy random variables, *Proceedings of the Royal Society of London* **407**:171–182.

M. L. Puri, and D. A. Ralescu, 1982, Integration on fuzzy sets, *Advances in Applied Mathematics* **3**:430–434.

M. L. Puri, and D. A. Ralescu, 1983, Strong law of large numbers for Banach space valued random sets, *Annals of Probability* **11**:222–224.

M. L. Puri, and D. A. Ralescu, 1983, Strong law of large numbers with respect to a set-valued probability measure, *Annals of Probability* **11**:1051–1054.

M. L. Puri, and D. A. Ralescu, 1985, Limit theorems for random compact sets in Banach space, *Mathematical Proceedings of the Cambridge Philosophical Society* **97**:151–158.

M. L. Puri, and D. A. Ralescu, 1986, Fuzzy random variables, *Journal of Mathematical Analysis and Applications* **114**:409–422.

D. A. Ralescu, 1979, A survey of the representation of fuzzy concepts and its applications, in *Advances in Fuzzy Set Theory and Applications*, M. Gupta, R. Ragade, and R. Yager (eds.), Amsterdam: North-Holland.

D. A. Ralescu, 1982, Toward a general theory of fuzzy variables, *Journal of Mathematical Analysis and Applications* **86**:176–193.

D. A. Ralescu, 1986, Radon-Nikodym theorem for fuzzy set-valued measures, in *Fuzzy Sets Theory and Applications*, A. Jones et al. (eds.), Amsterdam: D. Reidel.

A. Ralescu, and D. A. Ralescu, 1984, Probability and fuzziness, *Information Sciences* **34**:85–92.

INDEX

INDEX

INDEX

Abduction, 71
Absorption, 81
Achache, A., 96
Achinstein, P., 76
Adaptive control, 19, 25
Aggregate production planning, 53
Algebra, 21
  Boolean, 81
  Heyting, 75
Algebraic theories, of fuzzy sets, 84
Alston, W., 22
Alternative set theory, 17, 96
Ampere, A. M., 24
Analysis
  Fourier, 21
  functional, 21
  qualitative, 21
Approximate reasoning, 75
Arbib, M. A., 19, 25, 42

Aristotelian logic, 3
Array, 35
Artificial intelligence, 71, 72
Assilian, S., 66
Associativity, 81, 100
Automata theory, 42
Automatic train operation, 55
Automation, 3

Balaceanu, C., 24
Banach space, 116
Bayesian formula, for fuzzy sets, 143
Bayesian rule, 63
Bayesian statistics, 73
Behavior, 23, 43
  intelligent, 6, 23
Bellman, R. E., 94
Bigelow, J., 23
Boden, M., 20

Bolzano–Cauchy condition, 96
Bonissone, P. P., 60, 70
Boolean algebra, 81
Borillo, P. J., 96

Carnap, R., 76
Casti, J., 18
Catastrophe, 18
Category, 43, 75, 84
Cauchy sequence of sets, 96, 101
Central limit theorem
    for fuzzy random variables, 121
    for random sets, 118
Certainty factors, 73
Characteristic function, 15, 23, 79
Charniak, E., 71
Clark, K. L., 14
Commutativity, 81, 100
Complement, 80
Complexity, 18, 28
Composition, 41
Concatenation, 4
Conceptualization, 3
Conditional expectation, 139
Conflict resolution, 72
Congruence, 4, 43
Connection, 16
Context dependency, 59
Continuum, 7, 23, 58
    psychological, 22
Control, 1, 14, 16, 20, 24
    adaptive, 19, 25
    fuzzy, 69
    linear multivariable, 44
    rotary cement kiln, 51
    science, 21
Controller, 3
Convergence almost surely, 117
Convex hull, 117
Credibility, 77
Crisp set, 16, 23, 80
Cybernetics, 2, 23, 24, 43

Data base, 9, 14
    expert, 8, 13
    relational, 40
Debreu integral, 104
Deduction, 71
Degree of existence, 38
DeMorgan's laws, 81
Diagram, 41, 85
Dialectical logic, 3, 46
Differential equation, 21
Dinola, A., 24
Distributive lattice, 82
Distributivity, 81
Draganescu, M., 24
Dynamics, 1
    nonlinear, 20
Dynamo, 45

Einstein, A., 20
Engineering, 27
Euclidean norm, 82
Euclidean space, 65
Evidence, 76
Expected value, 100, 104
Expert data base, 8, 13
Expert system, 6, 24, 72, 75
Extension principle, 93

Feedback, 2, 18, 24, 25, 44
FORTRAN, 45
Fourier analysis, 21
Fuentes, R., 96
Function, 15
Functional analysis, 21
Functor, 87
Fuzzy algorithm, 70
Fuzzy arithmetic, 36, 92
Fuzzy conditional statement, 40
Fuzzy control, 69
Fuzzy convex set, 120
Fuzzy martingales, 141

Fuzzy number, 35, 38
Fuzzy random variable, 65, 99, 107, 110
Fuzzy relation, 41, 84
Fuzzy sets, 15, 23, 24, 79
  addition of, 92
  algebraic theories of, 84
  Bayesian formula for, 143
  integration on, 149
  theory, 79
Fuzzy set-valued measure, 125, 130
Fuzzy set-valued probability measure, 133
Fuzzy statistics, 61, 149

Gaines, B., 66, 70
Galois conexions, 96
Gaussian variables, 118, 122
Generalized set, 15, 93. *See also* Topos
Giertz, M., 94
Giuculescu, A., 24
Goal, 3
Golu, P., 24
Goodman, R. I., 149
Gordon, G., 44
GPSS, 44
Graded membership, 79
Gupta, M., 24, 66, 67, 94

Hajek, P., 96
Hausdorff distance, 101
Hegelian approach, to scientific truth, 46
Heyting algebra, 75
Hilbert's hotel, 17
Höhle, U., 95
Holmblad, L. P., 67
Holt-Modigliani-Muth-Simon paint factory data, 70

Human expert, 11
Human systems management, 23, 24

Identity, 3, 15
If-then rule, 41, 51. *See also* Production rule
Image, 83
Independence, 145
Inference mechanism, 6, 8, 61, 71
Information, 24
Integration on fuzzy sets, 149
Intelligent behavior, 6, 23
Internalization
  of a knowledge base, 4, 6
  of logic, 16
  of a model, 3, 44, 61
  of a semantic system, 8
Intersection, 23, 80
Invariance, 28
Inverse image, 83
Investment decision, 58

Kalman, R., 24
Kantian approach, to scientific truth, 46
Kelemen, M., 44
Kickert, W. J., 66
King, J., 66
Klement, E. P., 121, 149
Knowledge
  acquisition, 8
  base, 10, 14
    internalization of, 4, 6
  based controller, 4
  engineer, 10
  engineering, 61
Kochen, M., 24

Language, 47

Lattice, 85
Law of large numbers, 119
Lawvere, 18
Lebesgue theorem, 114, 129, 133
Leibnitzian approach, to scientific truth, 46
Lembessis, E., 66
Level set, 16, 37, 89
Limit theorems, 115, 149
Linear multivariable control, 44
Linguistic controller, 61
Linguistic model, 15, 21
Linguistic strategy, 51, 66
Linguistic value, 2, 21, 52
Linked-list, 37
Lockean approach, to scientific truth, 46
Logic, 16, 76, 94
  internalization of, 16
  multivalued, 18
  operators, 16
  symbolic, 15

McDale, F. G., 14
McDermott, D., 71
Machine intelligence, 71
Mamdani, E. H., 66
Management
  applications of system theory, 21, 43
  information systems, 12
  of uncertainties, 73
Mathematical programming, 21
Membership, 79, 80
  degree of, 88
Metric, 110
Metric space, 116
Minkowski addition, 100
Minski, M., 25

Model(s), 2, 14, 19, 27
  arithmomorphic, 28
  base, 9
  dynamic, 39
  internalization of, 3, 44, 61
  qualitative, 40
  simulation, 27
  uncertainty, 149
Modus ponens, 71, 73
Moore's family, 96
Moray, N., 19
Morphism, 85
Multivalued logic, 18
Myhill realization, 43
Myiamoto, S., 56, 68

Negoita, C. V., 21, 24, 43, 44, 48, 75, 94, 95
Nerode equivalence, 43
Nguyen, H. T., 149
Normality, 35, 65
Novak, V., 18, 97

Observability, 4
Odobleja, S., 24
Operators, 94
Optimization, 20, 95
Ostergaard, J. J., 66, 67

Padulo, L., 42
Parik's feasibility, 97
Plato, 24
Plausible reasoning, 75
Ponasse, 18
Population growth, 30
Power set, 79, 86
Precision, 16

Predicate, 15
Predicate calculus, 71, 76
Predictive approach, 56
Probability, 63, 72, 76, 125
  space, 102
Production-inventory system, 53
Production planning, aggregate, 53
Production rule
  aggregation of, 5
  chaining of, 6
  selection of, 5
Program, 9, 30, 31, 37
Prolog, 14
Proper classes, 17
Pullback, 4, 24, 34, 48, 94
Puri, M. L., 121, 149, 150
Purpose, 2, 23

Qualitative analysis, 21
Qualitative models, 40

Radon-Nikodym theorem, 138, 150
Ralescu, A., 150
Ralescu, D. A., 43, 74, 77, 121, 149, 150
Random set, 102
Random variable, 99
Ratiu, T., 43
Realization, 43
Reasoning, 75
Record, 37
Relational data base, 40
Rinks, D. B., 69
Rissland, E. L., 19, 25
Robustness, 21, 33
Rosenblueth, A., 23
Rotary cement kiln, control of, 51

Sanchez, E., 24, 67, 95
Saridis, G., 66
Sasaki, M., 95
Schroedinger's equations, 20
Science, 27
Selector, 104
Selfregulation, 23
Selfridge, O., 19, 25
Semantic approach, 2, 6
Semantic system, 7, 10, 11
  internalization of, 8
Semisets, 96
Separability, 119
Set theory, 15
  alternative, 17, 96
  Bolzano's, 96
  Cantor's, 17, 96
Shannon, 24
Sheaves, 75
Shapley-Folkman lema, 121, 123
Similarity relation, 4, 84
Simulation, 1, 29
  models, 27
Skala, H., 96
Sochor, A., 96
State, 1, 20, 39, 42
State equation, 2, 54
Statistics, 63, 115
Stefanescu, A., 95
Strong law of large numbers, 119, 149
Structural stability, 34, 44
Sugeno, M., 69, 95
Suitability, 70
Support, 37, 128
Sutherland, J., 46
Syllogism, 62. *See also* Inference mechanism
Symbolic logic, 15
Symscript, 45
Synthetic construct, 46

System(s)
  analysis, 8
  complexity, 18, 45
  expert, 6, 24, 72, 75
  ill-defined, 19, 25
  large scale, 18, 45, 48
  linear, 42
  multilevel, 21
  multivariable, 20
  nondeterministic, 43
  nonlinear, 21
  production-inventory, 53
  semantic, 7, 10, 11
    internalization of, 8
  societal, 46
  stochastic, 43
  simulation, 44
  theory, 42
    management applications of, 21, 43

Technique, 27
Teleology, 3, 23
Termini, S., 96
Theorem of representation, 18, 23, 37, 66, 91, 95, 132
Tocqueville, 48
Tong, R. M., 60, 66, 70
Topos, 18, 75, 93. *See also* Generalized set
Train, automatic operation of, 55

Transition, 42
Trillas, E., 96
Tyler, S., 47

Uncertainties, management of, 73
Uncertainty models, 149
Union, 23, 80
Unit interval, 15

Vagueness, 7, 22, 23, 28, 96
Van Nauta Lemke, H., 66
Vemuri, V., 45
Ventre, A., 24
Vopenka, P., 17, 96

Wenstop, F., 21
Wiener, N., 23, 24
Wohnam, W. M., 44

Yasunobu, S., 56, 68
Young, R., 20

Zadeh, L. A., 23, 24, 66, 70, 73, 94
Zimmermann, H. J., 66, 70

RAYMOND H. FOGLER LIBRARY
DATE DUE

BOOKS ARE SUBJECT TO
RECALL AFTER TWO WEEKS

OCT 3 0 1987

DEC 1 9 1988